"The story of geology is here told by recognized writers in the field, each contributing a word on one aspect of the subject. To give the work a certain unity, each selection is presented under a descriptive title appropriate to the plan of the book. The title of the work from which it is taken is given in the introduction to the selection."
From the Introduction.

Karen Lewis
3701 Crescent View Ave.
Duluth, Minnesota 55804

McGRAW-HILL PAPERBACKS
IN SCIENCE, MATHEMATICS AND ENGINEERING

Philip L. Alger, Mathematics for Science and Engineering	$2.95
D. N. Allen, Relaxation Methods in Engineering and Science	2.95
H. V. Anderson, Chemical Calculations	2.75
E. F. Beckenbach, Modern Mathematics for the Engineer	3.45
E. T. Bell, Mathematics, Queen and Servant of Science	2.65
W. F. Cottrell, Energy and Society	2.95
H. Cross and R. C. Goodpasture, Engineers and Ivory Towers	1.50
F. Daniels, Mathematical Preparation for Physical Chemistry	2.50
Martin Gardner, Logic Machines and Diagrams	2.25
Gerald Goertzel and Nunzio Tralli, Some Mathematical Method of Physics	2.45
Harriet Griffin, Elementary Theory of Numbers	2.45
Lyman M. Kells, Willis F. Kern and James R. Bland, Log and Trig Tables	.95
Paul E. Machovina, A Manual for the Slide Rule	.95
Henry Margenau, The Nature of Physical Reality	2.95
George R. Stibitz and Jules A. Larrivee, Mathematics and Computers	2.75
J. V. Uspensky, Introduction to Mathematical Probability	2.95
J. V. Uspensky, Theory of Equations	2.95
E. Bright Wilson, Jr., An Introduction to Scienfic Research	2.95
H. D. Young, Statistical Treatment of Experimental Data	2.95

Prices subject to change without notice.

The World of Geology

edited by
L. Don Leet
Professor of Geology, Harvard University
Florence J. Leet

McGRAW-HILL BOOK COMPANY, INC.
NEW YORK   TORONTO   LONDON

THE WORLD OF GEOLOGY. Copyright © 1961 by the McGraw-Hill Book Company, Inc. Printed in the United States of America. All rights reserved. This book, or parts thereof, may not be reproduced in any form without permission of the publishers.

Library of Congress Catalog Card Number:
61-9328
III

ACKNOWLEDGMENTS

The kind permission of the following individuals and organizations for use of copyrighted material is hereby gratefully acknowledged:

Scientific American, Inc., to adapt "Geology" by Reginald A. Daly, which appeared September, 1950, in *Scientific American*, copyright 1950 by Scientific American, Inc.; "The Dust Cloud Hypothesis" by Fred L. Whipple, which appeared May, 1948, in *Scientific American*, copyright 1948 by Scientific American, Inc.; "The Origin of the Atmosphere" by Helmut Landsberg, which appeared August, 1953, in *Scientific American*, copyright 1953 by Scientific American, Inc.; and to reprint "Tsunami" by Joseph Bernstein, which appeared August, 1954, in *Scientific American*, copyright 1954 by Scientific American, Inc.

Harvard University Press, to adapt pages 3-11 of *Prehistoric Life* by Percy E. Raymond, copyright 1947 by President and Fellows of Harvard College.

Yale University Press, to adapt pages 14-21 of *The Meaning of Evolution* by George Gaylord Simpson, copyright 1949 by Yale University Press.

John Wiley & Sons, Inc., to adapt pages 5-13 of *Evolution of the Vertebrates* by Edwin H. Colbert, copyright 1955 by John Wiley & Sons, Inc.

Rachel L. Carson, Oxford University Press, Inc., and Staples Press, to quote selections from *The Sea Around Us* by Rachel L. Carson, copyright 1950, 1951, by Rachel L. Carson; reprinted by permission of Oxford University Press, Inc., and Staples Press.

Life Magazine, to reprint from "Our 7-mile Dive to Bottom" by Don Walsh, which appeared February 15, 1960, in *Life;* courtesy Life Magazine, copyright 1960, Time, Inc.

Doubleday & Company, Inc., to reprint "Agassiz of the Ice Age" from the book *Giants of Geology* by Carroll Lane Fenton and Mildred Adams Fenton, copyright 1945, 1952, by Carroll Lane Fenton and Mildred Adams Fenton; and to adapt page 46 of *Our Amazing Earth* by Carroll Lane Fenton, copyright 1938 by Doubleday & Company, Inc.; reprinted by permission of Doubleday & Company, Inc.

McGraw-Hill Book Company, Inc., to adapt pages 11-88 of *Causes of Catastrophe* by L. Don Leet, copyright 1948 by McGraw-Hill Book Company, Inc.; and pages 17-30 of *How to Know the Minerals and Rocks* by Richard M. Pearl, copyright 1955 by Richard M. Pearl.

American Geological Institute, to adapt "The 1959 Eruption of Kilauea" by Jerry P. Eaton and Donald H. Richter, which appeared May-June, 1960, in *Geotimes,* copyright 1960 by American Geological Institute.

Howel Williams and University of Oregon Press, to adapt pages 1-12 of *Ancient Volcanoes of Oregon* by Howel Williams, copyright 1953 by University of Oregon.

E. P. Dutton & Co., Inc., to adapt pages 17-22 of the book *Earth Lore: Geology Without Jargon* by S. J. Shand, copyright 1938 by E. P. Dutton & Co., Inc.; adapted by permission of the publishers.

Harper and Brothers, to adapt pages 177-196 of *A Textbook of Geology* by Robert M. Garrels, copyright 1951 by Harper and Brothers.

Appalachia, to adapt "Origin of the Appalachian Highlands" by Marland P. Billings and Charles R. Williams, which appeared June, 1932, in *Appalachia,* copyright 1932 by Appalachian Mountain Club.

CONTENTS

Introduction *by L. Don Leet and Florence J. Leet*   *1*

Reginald A. Daly
GEOLOGY, 1900–1950   *13*
Adapted from "Geology"

Fred L. Whipple
THE ORIGIN OF THE EARTH   *21*
Adapted from "The Dust Cloud Hypothesis"

Percy E. Raymond
RECORDS OF LIFE   *31*
Adapted from *Prehistoric Life*

George Gaylord Simpson
THE DEVELOPMENT OF LIFE   *41*
Adapted from *The Meaning of Evolution*

Edwin H. Colbert
MAN AND GEOLOGIC TIME   *49*
Adapted from *Evolution of the Vertebrates*

Helmut E. Landsberg
THE ORIGIN OF THE ATMOSPHERE   *59*
Adapted from "The Origin of the Atmosphere"

Rachel L. Carson
OCEANS AND THEIR HIDDEN LANDS 67
Selected from *The Sea Around Us*

Don Walsh
THE OCEAN'S DEEP 91
From "Our 7-mile Dive to Bottom"

Carroll Lane Fenton and Mildred Adams Fenton
A GIANT OF GEOLOGY: AGASSIZ 103
From *Giants of Geology*

Louis Agassiz
ICE ON THE LAND 119
From *Essays*

Rachel L. Carson
ICE AND OCEAN LEVELS 131
Selected from *The Sea Around Us*

Joseph Bernstein
GIANT WAVES 139
"Tsunamis"

L. Don Leet
THE RESTLESS EARTH AND ITS INTERIOR 149
Adapted from *Causes of Catastrophe*

Jerry P. Eaton and Donald H. Richter
A VOLCANO ERUPTS 167
Adapted from "The 1959 Eruption of Kilauea"

Howel Williams
VOLCANOES BUILD THE LAND 177
Adapted from *The Ancient Volcanoes of Oregon*

S. J. Shand
WEARING AWAY OF THE LAND 197
Adapted from *Earth Lore: Geology Without Jargon*

Robert M. Garrels
WATER UNDER THE GROUND 207
From *A Textbook of Geology*

MATERIALS OF THE EARTH 219
Based on Carroll Lane Fenton, *Our Amazing Earth*
C. S. Hurlbut, Jr., *Mineralogy and Some of Its Applications*
Richard M. Pearl, *How to Know the Minerals and Rocks*

Marland P. Billings and Charles R. Williams
MOUNTAIN STORY 243
Adapted from "Origin of the Appalachian Highlands"

L. Don Leet and Florence J. Leet

INTRODUCTION

Geology, the science of the earth, tells us about the world, what it is made of, its age, its aging processes, its landforms, and its abyssal ocean depths. It traces life from the first sea-spawned vegetation and animals, through dinosaurs, to man—all on a majestic time scale that staggers the imagination.

These secrets of the earth are probed in many ways. Waves from earthquakes travel through the interior to bring out messages about its structure that are written on seismographs. Chemical and X-ray analyses reveal the composition of rocks and minerals. Mapping of sands, gravels, and boulders shows where glaciers once moved over the land. Examination of ancient caves and strand lines proves that the oceans have been hundreds of feet deeper than they are today, while submarine canyons and wave-planed volcanic peaks under deep water present the possibility that the oceans have also been hundreds, even thousands, of feet shallower.

Geology is all this, and much more. It is a synthesis of the natural sciences: astronomy, biology, chemistry, mathematics, and physics. And, above all, it has something for everyone. Who can experience or even hear about an earthquake or volcanic eruption without wondering about its cause? If you found a sea shell or

One of the highest mountains in the world, Kanchenjunga towers over 28,000 feet into the air among the spectacular ranges of the Himalayas. These mountains are composed of rocks that formed beneath the sea. (Vittorio Sella.)

fish solidly encased in the rock of an inland stream bed, or of a high mountain, would you wonder why it was there? Have you ever pondered the jumbled varicolored rocks or multitudinous grains of sand of a shoreline, the goldlike glitter of yellow mica in a piece of field stone, or the smooth symmetry of a quartz crystal? If these or any of the thousand and one phenomena all around us have stimulated so much as a fleeting question in your mind, you have peeked through a door into the world of geology. Anyone can walk through such a door and find treasures limited only by the dimensions of his curiosity and enthusiasm.

Our world of rocks, water, and atmosphere is composed of elements which were created at a definite time in the past. And it is not now as it was then; it is ever-changing. The mountains of today were sea bottoms in an earlier time; cycle has followed cycle of inner forces building up the land and external forces wearing it away. Oceans have invaded continents; mile-thick glaciers have overridden great areas, trapping water from shallowing oceans, only to waste away, have tropical vegetation take their place, and refilled oceans flood the shores.

For some centuries before Christ, Greek wise men filled volumes with poetry, drama, history, philosophy, logic, and metaphysics; some dealt casually with plants and animals; but treatments of topics now included under geology were for the most part unknown. One man of early Greece, the philosopher and teacher Aristotle (384–322 B.C.), did more than any individual in history to delay the birth of geology. He was the incarnation of the arbitrary dictum, unhampered by observation, experiments, or proof. He was incredibly persuasive in his day and, through his writings, in succeeding centuries, but he was uniformly wrong in the field of geology. Yet to this day, nearly twenty-three centuries after his death, there persists talk of "earthquake weather," an Aristotelian invention.

According to Aristotle, all things in nature were composed of fire, air, earth, and water. He explained rocks as formed under an undefined "influence" of sun and stars. He stated that the earth is a mere shell surrounding central fires, and that volcanoes are safety valves through which excess heat from these fires is occasionally vented. He stated that earthquakes are caused by an excess of air crowding into earth locally—leaving the atmosphere deficient and causing humid, stifling "earthquake weather"—where it is agitated by the central fires until, in escaping violently, it shakes the earth in an earthquake.

Shakespeare had Aristotle at his elbow when, in *Henry IV*, he had Hotspur say:

> Diseased nature oftentimes breaks forth
> In strange eruptions; oft the teeming earth
> Is with a kind of colic pinch'd and vex'd
> By the imprisoning of unruly wind
> Within her womb; which, for enlargement striving,
> Shakes the old beldam earth, and topples down
> Steeples and moss-grown towers.

Aristotle knew about fish remains encased in rock high above present sea level, but in his treatise *De Respiratione* had only this to say in explanation: "A great many fishes live in the earth motionless and are found when excavations are made," indicating that he thought the fish lived and died in the rock in which they were found.

Finally, in the mid-sixteenth century, Aristotle's authority in the field of geology was challenged by a German named Bauer, writing under the pen name of Agricola. He ridiculed the views of Aristotle that common stones, gems, and metals originate from some action of the stars, and set forth his own ideas based on personal observation of the earth's crust during residence in the mining districts of Saxony. He recognized and explained quite well the class of rocks we now call sedimentary, though he knew little about igneous and metamorphic rocks, for the chemical and physical data needed to study them were not available in his time. But, primarily, he broke the fetters of Aristotelian reliance on mystical guesses and worked from the ground up by assembling careful observations, then explaining. He had the good judgment to postpone speculation on remote ultimate causes, and confined himself to immediate processes. Appreciation of ultimate causes had to await recognition of atoms and their properties centuries later.

Modern geology was born in 1788 when a Scotsman, James Hutton, announced that "the present is the key to the past," meaning that processes such as weathering, erosion, vulcanism, and sedimentation observed and measured today have operated similarly in the past.

At about this same time, William "Strata" Smith in England and Georges Cuvier and Alexandre Brongniart in France learned that unique groups of fossils characterize the rocks of a particular time and can be used to correlate rocks from place to place. This discovery soon led to a realization that there was a progressive change in the forms of living things, change to which Charles Darwin in 1859 addressed himself in *On the Origin of Species*. The discovery of radioactivity by the Curies in 1898 started development of another major tool used in bringing geology to where we find it today. This discovery resulted in one of the great contributions to our civilization from the world of geology, the concept of geologic time.

According to Hindu chronology, the earth is nearly 2 billion years old. At the other extreme, James Ussher, an Irish bishop who specialized in Biblical chronology, in the seventeenth century announced that the world was created in the year 4004 B.C.; the only uncertainty was as to whether it was in the spring or the fall of the year. For many people, it was heresy to believe otherwise, and this attitude delayed further investigation of the problem. Then Hutton, using observed thicknesses of sedimentary strata and present rates of sedimentation, demonstrated that erosion and sedimentation must have been operating through an interval of time that could only be described as inconceivably long. He made no effort to estimate how long, but did report that he could find "no vestige of a beginning." Thereupon he was declared by many of his contemporaries to have denied creation. Lord Kelvin (1824–1907), world-famed British physicist, arguing from the premise that the earth has been cooling steadily since it solidified from a molten state in its formation, in 1862 announced that the earth is probably no older than 400 million years or younger than 20 million years. He later narrowed the limits by setting the upper figure at 40 million years, and at the beginning of the present century anyone who claimed an age of more than 100 million years for the earth was considered rash, to say the least.

The whole picture changed almost overnight with the discovery of radioactivity and the presence in rocks of measurable amounts of such elements as uranium, thorium, and potassium, which have been disintegrating at a steady rate since their formation. As these radioactive elements decay they change to other elements at rates which are precisely determined and constant, so that by careful measurement of the proportions of original elements and the by-products of their decay, we can now state how long the decay has been going on—that is, the length of time since the rock solidified. This method is now being used to date the formation of many rocks, as well as to determine the age of remains of plants and animals buried in them.

No rocks have been found that are older than 3 billion years. The earth is of necessity older than the oldest rocks found on its surface, and estimates of its age are around 5 billion years.

The elements which are constantly decreasing in abundance were not always here. They were created at a definite time in the past, as were all the elements of our earth and the solar system.

In the pages that follow, you will learn that 92 elements comprise the materials and life of the earth. Rocks are formed from a handful of these elements by processes neither mysterious nor unusual. There are no fires in the interior, which is solid to a depth of 1,800 miles, and though we know much about volcanoes and what they are not, we still have not proved satisfactorily why and how they operate. Earthquakes result when the solid earth is distorted to the breaking point as it writhes under forces continually at work to produce mountain heights and ocean deeps. Fossil fish, like all fish, lived and died in water, and their bodies were buried in sediment of the bottom which later hardened to rock and was elevated to heights at which it is now found.

Since necessities and luxuries of life alike come from the

earth and since geology is a key to finding them, there are many profitable as well as exciting professional careers associated with the finding. Among them are geochemistry, geomorphology, geophysics, mineralogy, paleontology, petrology, and seismology.

Geochemistry, literally "chemistry of the earth," broadly defined includes all parts of geology that involve chemical changes. There has been a tendency for geochemists to specialize particularly in study of the relative and absolute abundance of the elements and their isotopes in the earth and in the distribution and movements of elements in various parts of the atmosphere, oceans, and crust. Geochemistry now plays an important role in solving problems connected with finding and mining metallic ores, gas, oil, and radioactive minerals.

Geomorphology, "form of the earth," is the branch of geology that treats of surface features of the globe, their form, nature, origin, and development, as well as the changes they are undergoing. This broad charter naturally embraces many things, which include the work of water in streams and elsewhere in shaping the surface by moving material from one place to another, occurrences and characteristics of underground water so essential to civilization, the breaking down of rocks to form soils and supply sediments to the sea, and the dating of archaeological sites.

Geophysics, "physics of the earth," applies principles of physics to studies of the earth's structure, composition, and development; the atmosphere; the oceans. Methods of geophysics are applied to geologic exploration or prospecting for gas, oil, and minerals, which occur in such a way that their finding can be expedited by determining the depths, form, and sometimes the nature of buried rocks.

Mineralogy, the science of minerals, deals with their crystal shapes, physical and chemical properties, classification, and identification. Since minerals are naturally occurring elements or compounds formed without the aid of life processes, they comprise most of the rocks and other materials of geology. One of the more glamorous aspects of mineralogy specializes in gems and precious stones. Most of geology involves mineralogy in one way or another.

Oceanography, the science of the oceans, consists of studies pertaining to the sea. It includes such subjects as ocean boundaries and bottom topography, the physics and chemistry of sea water, currents, and many phases of marine biology. It is a branch of geology that has grown very rapidly in recent years because of the many government and private interests concerned with the results of its investigations.

Paleontology, the science of ancient life, deals with animal and plant life of past geologic ages, and is based on the study of fossils. It is applied commercially to identify rock layers through which drills will go in search of oil and gas.

Petrology, the science of rocks, is the study of the natural history of rocks, including their origin, present condition, changes, and decay. It is, therefore, basic to many phases of geology, and has an application in practical problems as well as in research.

Seismology, the science of earthquakes, is concerned with the cause, distribution, and effects of earthquakes; with instruments for recording earthquake waves in the ground; and with the interpretation of seismographic records to learn about the structure of the earth's crust and interior. It has been employed extensively in mapping structures favorable to the accumulation of gas and oil as well as structures of interest for other reasons. It is now being used to develop methods for detecting underground nuclear tests. There are several hundred seismograph stations throughout the world.

As geology pushes into the frontiers of knowledge, there is an ever-increasing need for teachers to transmit knowledge to others. This need is becoming critical as the realization spreads that geology is the best avenue for developing elementary and high school as well as college understanding of all sciences, since it is a tangible, interesting application of them all to the world in which we live.

Above all, geology offers to everyone an understanding that

underlies the fullest enjoyment of the world around us. To the awesome beauty of mountain scenery it adds the dramatic story of how this emerged from sea bottom muds through the alchemy of time. To the wonders of plant and animal life it adds the fascinating details of their ancestry, also traced to the sea. It gives character and individuality, a sense of unity and purpose, to the loam of a backyard garden, the sands and gravels of a stream bed, the rocky ledges of a road cut. Emerson said, "The unequivocal mark of distinction is an ability to see the miraculous in the common." In no way can this ability be nurtured better than through geology. If a reader is led to look at the world through new eyes, to discover the exhilaration that comes from what can be seen, the purpose of this book will have been fulfilled.

The story of geology is here told by recognized writers in the field, each contributing a word on one aspect of the subject. To give the work a certain unity, each selection is presented under a descriptive title appropriate to the plan of the book. The title of the work from which it is taken is given in the introduction to the selection.

Much of our present knowledge of geology has been acquired during the twentieth century. At the mid-point of the century, **Scientific American** asked leaders in various branches of science to review developments in their fields during the previous fifty years. For geology, this was done by Reginald A. Daly, professor at Harvard University from 1912 to 1942, who had himself contributed notably to its progress.

This half century saw the birth of the concept of geologic time, measured by the decay of radioactive elements; realization that thick sheets of glacial ice bend the earth's crust downward from 1,000 to 2,000 feet; recognition of the source of heat generated in rocks; and the opening to exploration from the air of great areas of the earth's surface previously inaccessible. Study was begun to determine the shape of the rocks of the globe covered by oceans, where features were found that are best explained by past changes of sea level on a grand scale.

The following selection, adapted from the article "Geology" by Reginald A. Daly, one of the outstanding pioneers of the period, gives us the flavor of past discovery as well as an awareness of vast realms yet to be explored.

Reginald A. Daly

GEOLOGY, 1900-1950

Geology's foundations were laid toward the end of the eighteenth century by a few lonely geniuses of western Europe. The scientific study of our planet naturally had to begin at the surface of the dry land. The pioneers in geology had only a few simple tools: hammer, compass, collecting bag, notebook, topographic map, and a good pair of legs. With such equipment, still indispensable to the field geologist today, they explored old sedimentary rock formations of western Europe, where streams cutting through steep slopes had exposed rock layers, and by simple inspection they deduced the relative ages of these deposits.

The early geologists soon discovered that in each successive group of rock layers there were entombed certain fossils, chiefly of sea animals and plants, which were characteristic of the time in which they lived. During the nineteenth century, geologists proved that this marine life had gone through an irreversible evolution. Thus the "guide" fossils of Europe became a kind of calendar for reckoning the relative antiquity of the principal sedimentary rocks, not only in Europe but on every continent and large island where similar fossils were found. They made it possible to compose a rough time scale for the infinitely varied events in the earth's history, in so far as these were registered in the accessible rocks.

In our century, with the advent of the automobile and the airplane, large sections of the earth's surface structure have been mapped. Sweeping geologic reconnaissance has been made of wide stretches of Africa, Australia, Canada, and even forbidding Antarctica, whose border reveals a belt of folded rock formations that appears to be part of the vast mountain belt rimming the Pacific—the greatest mountain system on the earth. It has also produced a number of more detailed local geologic studies.

A truly scientific history of any phenomenon must explain it in terms of its origin; the geologist's problem here is complicated by the fact that the origins of continents, islands, ocean basins, land basins, mountain ranges, high plateaus, volcanoes, and mineral deposits are hidden in the depths of time as well as in the depths of the earth. The geologist cannot get far in this search without help, so he has harnessed himself into a team with the geophysicist, the geochemist, and the astrophysicist. The partnership is responsible for the major achievements of the first half of the twentieth century.

We delve first into the question of the earth's time scale. The geologists of the nineteenth century demonstrated that our planet had a long history of leisurely change, with many successive cycles of erosion, sedimentation, and mountain building. They divided this history into four major eras: the Pre-Cambrian, the Paleozoic (ancient life), the Mesozoic (middle life), and the Cenozoic (recent life). Their estimates of the minimum age of the earth ran to hundreds of millions of years, but they had no means of making anything like a precise measurement either of the total age or of the length of the successive eras. In this century, the physicist and the geochemist have provided geology with a remarkably handy clock for this purpose: the radioactivity of rocks.

This time scale is now one of the chief mental tools of the student of earth history. And the discovery of radioactivity led on to another finding of fundamental concern to geology. This has to do with the heat of the earth.

Before radioactivity was known, geologists had no reason to

doubt that through all its history the earth had steadily been cooling off from an original molten state. But as studies of radioactive elements were pursued into deep mines and tunnels, it soon became evident that considerable heat must be created in the earth by the spontaneous disintegration of atoms. The earth, in short, is a true furnace.

This discovery opened a world of new exploration, for it has a most important bearing on the problem of what forces and processes have molded the earth's main structural features. Since heat must play a key role in these processes, obviously our ability to find out what they are depends heavily on the investigation and correct interpretation of the temperature conditions down through various levels of the crust and the underlying material on which it rests.

During this century, the temperature of the upper part of the earth's crust has been measured at hundreds of deep boreholes and mines in Europe, North America, and South Africa. All the measurements show that the temperature increases with depth. But there are good reasons for questioning whether the rate of this increase is the same at great depth as it is near the surface. One reason is some uncertainty about the distribution of radioactivity in the body of the planet. Another is uncertainty about how hot the infant earth was at the beginning. A third is uncertainty as to the possible effects on surface temperatures of localized convection currents that might have brought up especially hot material from great depth. Here, then, the geophysicist is in trouble, and he will do well to consider some relevant facts won by the geologists.

A vital discovery relating to the temperature of the material just below the earth's crust was made by geologists when they studied the reactions of the earth's body to heavy and extensive loads piled on its surface. Just such loads are represented by the existing Greenland and Antarctica icecaps. With seismographs, geologists proved that these caps are thousands of feet thick. Other workers made it clear that the glacial icecaps that lay on northwestern Europe and northeastern North America during the ice ages of the Pleistocene period must have had respective maximum thicknesses of about 9,000 and 12,000 feet. They showed that under these huge loads the crust of the earth was bent down as much as 1,000 feet in Europe and nearly 2,000 feet in America and, further, that these basinings of the crust continued even after the vertical excess pressure on the bottom of the crust had become small. This result of their field work shows that the material at and below the depth of a few tens of miles is extremely weak. The only good explanation of that weakness is that the material is made plastic by high temperatures.

The cooperative studies of geologists and geophysicists have also established a convincing explanation of the reason why the continents and large islands stand an average of 3 miles above the ocean floor. The dry-land parts of the earth's crust, being predominantly granitic, are lighter than the crust beneath the open ocean. In consequence, the continents are "floating" on the plastic layer which underlies the whole of the crust.

A related problem is the origin of the lava type of rocks, which have risen to the surface of the earth, and are still rising, as hot, liquid eruptions from the depths below. Geologists of the nineteenth century and early part of the twentieth thought this could be explained in a comparatively simple way. They conceived the planet to have a crust of crystallized rock resting on a substratum of molten, gas-charged basalt; from time to time molten material from the substratum erupted through fissures in the crust, making piles of basaltic lavas and other types of melts that were the products of chemical reactions between the liquid basalt and the invaded rocks. But doubt has been cast on this theory by recent geophysical evidence which makes it appear extremely improbable that any completely molten shell now exists near the surface of the earth. Some authorities have suggested that the lavas, including those of volcanoes, come from isolated pockets of hot liquid in the thick, solid crust.

Exploration of the ocean depths by echo-sounding methods developed during this century has added several new mysteries. One of these was raised by the discovery that the floor of the

Pacific Ocean is ornamented with many "sea mounts"—tall, conic hills whose tops are from a few hundred fathoms to fifteen hundred fathoms below the sea surface. They appear to be volcanic piles, but their tops have been beveled flat. It is not yet clear how and when they were so truncated.

A second mystery relates to the discovery of the surprising ruggedness of the continental slopes—the plunging undersea banks that are found off the shores of the continents. Geologists of the nineteenth century assumed, on the basis of sketchy soundings, that these slopes were smooth deposits built out by prolonged erosion from the continents. But this proved to be a serious error; more extensive and more accurate soundings made by the new methods in this century have shown that the slopes are cut by deep furrows like river valleys, now called submarine canyons. How these remarkable fretworks on the flanks of the Americas, Europe, Asia, and Africa developed is a matter of considerable debate. They seem to have been formed during the past million years, and some geologists suggest that they were actually the sites of rivers during the long ice-age intervals when sea level was lowered by the piling up of ice on the land.

This matter of the rise and fall of sea level has been a subject of fruitful controversy among geologists. It may hold the answer not only to such questions as the origin of the submarine canyons and of the living coral atolls and barrier reefs of the Pacific, but also to the broader problem of the stability of the earth's crust throughout the whole tropical belt of the planet during the last million years. Indeed, the history of the Pleistocene's ice ages has become a matter of great interest to biologists, anthropologists, and geographers, as well as geologists. The investigations of the past half century have yielded convincing proof that the icecap in the Northern Hemisphere made a series of advances and retreats during the Pleistocene, producing alternating periods of cold and warm climate, and the same sequence seems to have taken place in the Southern Hemisphere. Oceanographers have found evidence that these climatic changes were world-wide; in cores taken from the floor of the deep ocean at widely differing latitudes, warm-water fossils alternate with cold-water species at successive levels. The computed duration of these alternate periods of warm and cold ocean temperatures matches well with the length of the glacial and interglacial periods on the lands.

Brief and incomplete as this review of the twentieth century has been, it illustrates the youth of geology as a true science. For every problem so far solved, many new ones have arisen. The working out of the true story of the earth's evolution obviously still lies in the distant future. It is clear that the answers will require experiments and investigations by physicists, chemists, seismologists, oceanographers, and geodesists as well as by geologists. Perhaps it is not too much to hope that ultimately the cosmogonists will be able to supply a crucial piece of knowledge on which the solution depends; namely, a reasonably assured description of what our planet was like in the beginning—its constitution and its temperature.

In the words of a famous teacher of geology, "There is a world of work to be done!"

Geologists, in learning about the earth, have never ceased to wonder about its beginning, about how and when it was formed. This is not mere idle speculation, for the question has a bearing on the solution of some of the major problems of geology: Is the liquid core left over from an originally molten globe that has been cooling for 5 billion years; or did a smaller, solid ball sweep up material from space and grow to its present size, then start melting from the heat of radioactive elements trapped inside? Did the rocks and minerals we find today have their start in a common melting pot, or were they ready-formed at the start? We do know that the world in the beginning was not as it is today; a record of billions of years of change has been read from the rocks. So how did it start?

Any theory of the origin of the earth must explain not only the kinds and ages of rocks, the structure of the interior, but also its relationship to the solar system and the universe. One early idea was that a great mass of whirling nebulous matter in space gradually condensed, now and then throwing off a ring that gathered itself together as a planet. The final stage was collapse of everything that was left, to form the sun. Another suggestion was that a passing star came so close to the sun—origin of which was unexplained—that its gravitative pull drew great bolts of matter from the sun. These had relatively small blobs among them that gradually grew in size by sweeping up material as they swung around the sun, until they attained the sizes of the planets. A modification of this idea attributed the less dense outer planets to material from the passing star—until Pluto, the most distant, was found to be as dense as the inner planets. Other suggestions have also been made, then discarded as it was found that they failed to explain new observations.

Latest among theories of the origin of the earth is one that incorporates something from each of the others and is called the dust-cloud hypothesis. This starts with finely divided material in space, a dust cloud, and explains how it could have gathered into larger and larger clusters, which finally collapsed into the planets and the sun. In 1948, Prof. Fred L. Whipple of Harvard University, Director of the Smithsonian Astrophysical Observatory in Cambridge, Massachusetts, and world-renowned authority on comets, meteorites, and upper-air phenomena, discussed the origin of the earth under the title "The Dust Cloud Hypothesis," in **Scientific American.** When we asked him for permission to reprint the article, he pointed out that he was reluctant because thinking in that area has been changing so rapidly that the hypothesis is already teetering and might be modified or supplanted any day. It still is the most successful one we have in explaining many baffling features, however, so we have compromised by editing the original article, while retaining all possible of its splendid clarity of thought and expression; substituting some new numbers where appropriate; and obtaining Professor Whipple's approval of the edited version only for what it is, an exposition of the dust-cloud hypothesis.

Fred L. Whipple

THE ORIGIN OF THE EARTH

The beginnings of our world are beclouded in the swirling mists of countless ages past. What process created the planets and the stars? Are new ones still being formed? Or were all that now exist made in one fell swoop? And if so, when did that happen? Scientists are making progress in their study of these fascinating questions.

The study begins with the earth itself. Its present condition tells little about how it was formed, but like an oldster's brittle bones, its crust does yield evidence of its age. This evidence consists of traces of the products of radioactive disintegration within the earth's oldest rocks. Over vast intervals of time, radioactive uranium in the rocks breaks down into other elements. In 5 billion years half of the atoms in a given amount disintegrate. The products are helium and a stable isotope of lead, both of which are trapped in uranium-bearing rocks. By careful measurement of these minute traces of helium and lead, the age of the rocks can be determined. Analysis has shown that the oldest earth rocks existed much as they are today for some 3 billion years.

But how long the earth had been in existence before these uranium-bearing rocks formed, we do not know. There is strong astronomical evidence, however, that it is not much older than 12 billion years. We believe that this is evidenced by the fact that the entire universe appears to be expanding in such a way that if you work the process backward, it would all have been concentrated in one small part of space about 12 billion years ago.

It is well to remember that this expansion concept may be wrong. Perhaps some fundamental error creeps into our calculations when we project our theories so far back in time from our present observations. We may be deceived in thinking that the universe is expanding. Perhaps we are observing some strange effect that space and ime exert upon light rays which have been traveling for hundreds of millions of years.

In any case, our problem is to explain how the spinning sphere on which we live came into being. Many possibilities have been explored. But one by one the old, familiar theories have been shown by astronomers to possess weaknesses that make them implausible. The latest theory also may be wrong, but it does seem to fit the facts as we know them better than the others.

Any theory about the evolution of our planetary system must explain certain striking characteristics: (1) the planets all move in the same direction and very nearly in the same plane as the

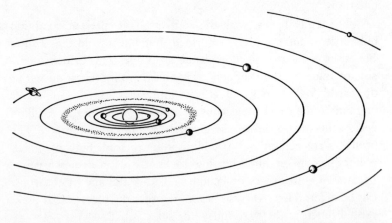

Our solar system today. Reading outward from the sun are the four inner planets: Mercury, Venus, Earth, and Mars; the asteroids; and the outer planets Jupiter, Saturn, Uranus, Neptune, and Pluto. In this drawing, the size of the planets has been exaggerated. Satellites, except for the rings of Saturn, have been omitted.

earth's orbit; (2) their orbital paths around the sun are nearly circular; (3) almost all the planets rotate on their axes in the same direction in which they revolve about the sun; (4) most of them have moons or satellites—Jupiter has 11—which usually revolve about the planet in the plane of its rotation and in the same direction.

Thus, the theory must account for a great deal of regularity in the system. But there are also some irregularities. Among these, Neptune's single satellite revolves backward, as compared with the rest of the solar system, although Neptune itself turns in the forward direction and is thus properly oriented. Some of the satellites of Jupiter and Saturn also are contrary.

The latest theory begins with the fact that interstellar space, formerly supposed to be empty, is now known to contain an astonishing amount of microscopic material. It has been calculated that the total mass of this interstellar dust and gas is as great as all the material in the stars themselves, including all possible planet systems. In other words, for every star there is an equal amount of dust and gas dispersed in space. The immensity of this quantity of material is beyond the grasp of human imagination. In the Milky Way alone, it comes to 100 billion sun masses. Yet interstellar space itself is so vast that the dust and gas are scattered more thinly than in the highest vacuum that can be created on earth.

We have a good deal of information about the composition of this nebulous star dust. The elements that we can detect are the ones with which we are familiar—hydrogen, helium, oxygen, nitrogen, carbon, and so on. These atoms slowly combine to form dust particles which in some places gather together into large clouds. While the structure of the dust particles is uncertain, it appears that most of them are very small—of the order of a fifty-thousandth of an inch in diameter. Evidence of their size and of the fact that they actually are dust particles is afforded by the way in which they scatter the light from distant stars, producing dark areas on a photographic plate. The small amount of starlight that filters through these dust clouds is reddened—for the same

Horsehead Nebula in Orion. This is a cloud of interstellar material similar to the one from which our solar system is believed to have formed. (Mount Wilson and Palomar Observatories.)

reason that the sun appears reddened during a dust storm: the long waves of red light are less scattered by small dust particles than are the shorter waves of other colors.

What collects these dust particles into clouds? It has been suggested that it might be the pressure of light. The pressure of light, which is so exceedingly small that it cannot ordinarily be observed, is nonetheless real enough. It is capable of forcing fine material away from the head of a comet, causing its tail to be pointed away from the sun at all times. As a comet comes into view from the darkness of outer reaches of the solar system, its tail trails behind it as seems normal for a fast-moving body. But as the comet's head circles the sun and starts away from it, the tail swings around, always keeping the head between it and the sun, and the whole system recedes tail first. Under rather unusual, but possible, circumstances the light from stars would tend to force interstellar dust into larger and larger clouds. In the starlight of space, each dust particle casts a shadow. After a few particles are collected into a small cloud, the cloud casts a larger shadow in the starlight, and this shadow falls on particles in its neighborhood. These particles are then drawn into the cloud, making it larger and larger. If such a cloud is not too much stirred by its motion through other banks of dust and gas, and if too bright a star does not pass through it and scatter the particles by its light pressure, the cloud will continue to draw in dust. Finally, it will attain a mass and density sufficient for gravity to become stronger than light pressure. The cloud will then begin to contract. Calculations show that for a dust cloud with the same mass of material as the sun, the two forces would be about equal when the diameter of the cloud was some 6,000 billion miles. This distance is sixty thousand times the distance of the earth from the sun. It has been further calculated that such a cloud might develop and collapse into a star in less than a billion years.

Let us consider our solar system as a case study of the formation of a star and its satellites. We have a huge dust cloud, as described above, which has begun to condense under gravity. There will be minor turbulent motions of the material with it—subclouds, or streams of dust—that slide by each other or collide. But these motions cannot all be in the same direction. In order to explain the present slow rotation of our sun, we must assume that the motions of the streams in the original dust cloud canceled each other and that the cloud as a whole did not rotate. All rotation that exists in the whole system in the beginning will remain there forever, and the rotation becomes more rapid if the system shrinks in size. Thus, if the great dust cloud from which the sun was formed had had any appreciable rotation to begin with, after its collapse the condensed sun would have rotated with great speed. But actually our sun turns very slowly; it takes nearly a month to make a complete rotation. Consequently the original dust cloud must have been almost stationary.

Having accounted for the sun's slow rotation, let us go back to the original dust cloud. Under the force of its own gravity, it has begun to condense. At first it collapses very slowly, because the motions of its internal currents and streams resist its contraction. A moving group of particles is harder to collect and compress than one which is standing still. But in the course of millions of years the random motions of the streams within the cloud are slowed down by collisions and friction. Meanwhile the cloud contracts more and more powerfully as it becomes smaller, because as its density increases, the force of gravity among the particles increases. The net result, with resistance diminishing and gravity increasing, is that the cloud collapses faster and faster. Its final collapse from a size equal to that of the solar system (the diameter of the orbit of Pluto) would require just a few hundred years. Due to the increased pressure in the contracting cloud, its temperature rises enormously. In the last white-hot phase of its collapse, the sun would begin to radiate as a star. Its central temperature, due to contraction, would become great enough to start the cycle of nuclear reactions among the carbon, hydrogen, and helium which keeps the sun radiating, but no detailed theoretical study of this phase has yet been made.

Now we must account for the evolution of the planets from the same great dust cloud. We return to the cloud before it has begun to shrink appreciably, and follow the largest stream in the cloud. If the dust in this stream is sufficiently dense, the stream condenses into minor clouds. They may be strung out in a series. As these clouds drift along together, roughly in the same direction, they will pick up material less compact than themselves; hence they will grow slowly, feeding on portions of the great cloud. As they grow, the minor clouds, now embryo planets or planet-clouds, begin to spiral slowly in toward the center of the main cloud. They have gained in mass but not in angular momentum, so they move toward the center of gravity. Some move in

THE ORIGIN OF THE EARTH

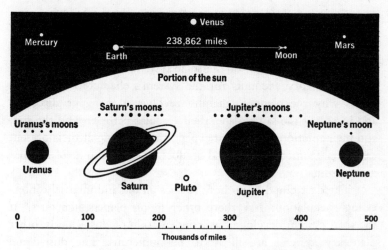

The size of the planetary bodies compared with a portion of the sun.

more rapidly than others, their rates depending on their sizes and on chance encounters with other streams.

If the great cloud remained spread out forever, all the embryo planets would eventually wind up at its center. But long before some of them have completed their spiral, the main cloud collapses and forms the sun. Its rapid final collapse leaves a number of embryo planets stranded in their orbits, outside the collapsing cloud. Some are trapped too near the center and are pulled in or destroyed in the sun's heat. Others are far enough away to remain intact. They condense and become planets. Some of them may be at enormous distances from the sun. For all we know, there may be planets in our system beyond Pluto, the farthest one that we can see. Even a great planet like Jupiter, the largest in our group, would almost certainly have escaped discovery if it had been at a distance from the sun one hundred or more times that of Pluto. Pluto itself, which is about the earth's size, is barely within the range of probable discovery.

When first formed, the planets are hot, perhaps hot enough to be in a molten condition. But since they are relatively small, their heat of contraction is not sufficient to start the nuclear reactions that would make them radiate permanently like stars. Gradually they cool off.

We have described, then, how the dust-cloud hypothesis accounts for the origin of the solar system. Now let us see how well our theory accounts for the system's characteristics. It explains why the planets generally revolve in the same direction and in nearly the same plane, their circular paths around the sun, and their rotation. It does not explain their spacing at their present distances from the sun. And it does not explain their angular momentum.

The dust-cloud hypothesis revives an old and intensely interesting speculation: Are there other living planets like ours? If the solar system developed by condensation of dust clouds, other planetary systems are likely to be numerous. The dust-cloud theory thus suggests a possibility that worlds with human or intelligent life may be fairly frequent throughout the universe.

The consideration of such questions no longer belongs only in the realm of science fiction. If intelligent beings exist on other planets, we may some day establish radio contact with them. Conceivably we may be able to send ships into space to cruise among planetary systems belonging to other stars in our neighborhood. There our descendants may find strange types of intelligent beings, people like ourselves, or at least settle the argument.

Much of our knowledge in geology is based on the study of records of life in the geologic past. These records are fossils, the remains or traces of animals or plants which have been preserved by natural causes in the earth's crust. They do not include organisms which have been buried since the beginning of historic time, that is, since man-made records in writing or inscriptions have been available.

That certain types of rock enclose fossils was known long ago. Ancients around the Mediterranean believed that fossil fish high above present sea level were formed by spontaneous generation, or were the remains of fish that had spawned from eggs accidentally misplaced, lived without moving, and died where they were found. It was when the actual processes of becoming a fossil were explained that geology was born. Fossil fish once lived in water, as all fish have, and when they died, they were covered by bottom sediments that later hardened to rock and were eventually elevated to their present positions.

"Strata" Smith, in the middle of the eighteenth century, observed that there were similar groups of fossils in rock layers many miles apart. He assumed that these groups lived at about the same time and therefore the rocks enclosing them were of the same age. He also noted that fossil plant and animal types were very different in the bottom and top layers of large masses of rock. Long intervals of time must have separated deposition of the first and last of these, he reasoned. From interpretation of these fossil records a picture of life and land changes throughout time began to emerge. Today, we find fossils carrying our knowledge back half a billion years in great detail, and beyond that sketchily.

Not all the fossils in the world have been found. In many places, yours could still be the first hands to touch and eyes to see a fern that lived 100 million years ago, or a trilobite, the three-lobed cousin of modern crabs and lobsters that has been extinct for 200 million years.

In the following selection, fossils are described and some of the problems of identification and interpretation are discussed. It is adapted from a book, **Prehistoric Life,** by the late Percy E. Raymond, former professor of paleontology at Harvard University, a great teacher who never lost sight of his pupils' interest and abilities.

Percy E. Raymond

RECORDS OF LIFE

Much has been written in recent years about the early history of the earth in so far as it can be deduced from astronomical and physical data. The evolution of the world would have been futile, however, had it not been for the introduction of life. As to how life originated, geology unfortunately gives little information, but that the earth has supported life for countless millions of years is clearly shown by the remains of animals and plants entombed within the sedimentary rocks during their accumulation and preserved to the present time. These remains serve a twofold purpose. Not only do they give a clue to the history of life upon the globe, but when properly studied and interpreted, they reveal much of its physical history. What prehistoric implements are to the archaeologist, or the inscriptions incised by ancient peoples upon enduring rock are to the historian, such are fossils to the geologist. Fortunately, their study is not nearly so difficult as that of artifacts or inscriptions, nor does it require so technical a training; yet it produces results of the same order of accuracy.

To study fossils, it is necessary to have some knowledge of living animals and plants, for fossils are more or less perfectly preserved remains of organisms, or evidences of their former existence. The simplest animals are microscopic creatures, found commonly in fresh water, which look like bits of transparent jelly. Some of them have no particular shape, or rather a constantly changing one. They possess no heads or bodies, arms or legs, eyes or mouths, or even any digestive tracts; in fact, they are organless, except for a spot of slightly denser material called a nucleus. A well-known example of such a creature is the amoeba, familiar to everyone who has looked through a microscope.

It may be convenient, if one lives in water and does not want to go anywhere in particular, to have a jellylike consistency, but most animals have some sort of device for stiffening the body. This strengthening matter, the skeleton, may be either external or internal. An external skeleton serves secondarily as a means of protection. The most common material of skeletons is calcium, either in the form of carbonate or phosphate or as a combination of the two, but it may be silica or hornlike chitin. Plants as well as animals have a stiffening material, the woody tissue, which, like chitin, is a carbohydrate, called cellulose. Silica, too, is present in some plants, giving their cutting power to grasses and sedges. Many of the aquatic plants (algae) secrete large amounts of calcium carbonate. Most organisms, in short, possess "hard" parts, bones, shells, spicules, or woody material. Were it not for these relatively indestructible portions there would be little record of the history of ancient life. "Soft" parts, the protoplasmic tissues, flesh, and cartilage, decompose very rapidly through the action of bacteria. Hard parts, although they decay, do so slowly; consequently there is a chance that they may be preserved as fossils.

What is the process of becoming a fossil? It is merely preservation, either by the checking of decomposition or by the replacement of the hard parts by some relatively durable substance. Anything unfavorable to the life of bacteria impedes decay. Very dry air, a low temperature, sea or bog water, burial in mud or volcanic ash, an encrustation of pitch, gum, or calcium carbonate, all have a more or less preservative effect, so that decomposition is either retarded or entirely prevented. When bacteria are entirely excluded not only the hard parts but the soft as well may be preserved as in a modern refrigerating plant. The most famous

Photograph of a fly that was trapped in gum oozing from a now extinct pine tree 60 million years ago. The gum hardened into amber, which has almost perfectly preserved the external skeleton of the insect, even to the tiny hairs on its legs. (Smithsonian Institution.)

instances of cold storage are those of the remains of mammoths and rhinoceroses occasionally found in the frozen gravels and ice of Siberia. Another case of remarkable preservation is that of insects in amber. While it was a sticky gum exuded from a species of pine, numerous insects were trapped in it, to be preserved as it hardened. Although amber of many ages is known, the most abundant insect-bearing material is found in rock layers about 25 million years old on the Prussian shores of the Baltic.

Suppose an organism to be buried in some substance which retards decay. Conceivably the sand, clay, or calcareous ooze which surrounds it may become sufficiently compacted while the organism retains its original form to hold its shape. Then subsequent decay of the object will leave a hollow mold. This may be preserved as such, in which case it would itself be a fossil, or percolating waters may eventually fill it with calcite, silica, pyrite, mud, or even sand.

There is another type of replacement in which decay of the skeleton occurs in the presence of mineralized waters, so that for each particle removed the water gives up a bit of its mineral matter, producing a delicate replica of the whole original structure. This process has produced the petrified wood now seen in the Petrified Forest of Arizona or other areas of badlands in the western states. Such petrified wood, although entirely changed to stone, still shows the characteristic rings of growth, knots, and other features of modern trees. This type of preservation can occur within the skeletons of animals, but is much more commonly found in the replacement of plants.

Some fossils are so preserved that they retain indications of the shape of the internal organs of animals, even though no tissues actually remain. Thus there are certain creatures which commonly ingest large quantities of mud with their food. The mud-filled alimentary tracts of a few such organisms have been recovered, showing the shape of stomach and intestines. The fossil excrement of animals also shows something of the shape of the alimentary canal but is particularly interesting because it often contains undigested remains of food. Impressions of skin are sometimes found, and, more rarely, impressions of other soft, perishable tissues. Even jellyfish and other equally delicate organisms have been found at a few places where the rock is of fine grain. The two most famous localities for such fossils are the 450-million-year-old rock layers above Burgess Pass, near Field, British Columbia, discovered and explored by C. D. Walcott, and the quarries in the 140-million-year-old lithographic limestone at Solenhofen, Bavaria. These latter have been worked for many years and have furnished some of the most remarkable and important of fossils, running the gamut of the animal kingdom from jellyfish to flying reptiles and toothed birds.

Artificial structures made by organisms are occasionally found. Some burrowing animals lined their habitations with bits of shells or sand which they cemented together, whereas others made nests which are occasionally preserved in the rocks. The most abundant artificial structures are the various implements of prehistoric man.

Fossils are the remains of organisms or the direct evidences

This rock fragment contains some fossils of trilobites, small marine animals which were the most abundant organisms 400 to 500 million years ago but became extinct 200 million years ago. (Smithsonian Institution.)

of their former existence, preserved in the earth's crust. This definition, although generally acceptable, has the fault of being rather too inclusive, since it makes no reference to the time of burial. As a matter of convenience, many paleontologists, including the author, arbitrarily exclude from the category of fossils all things which have been buried since the beginning of historic time. Such a course helps the paleontologist to avoid duplicating the work of the archaeologist, botanist, and systematic zoologist, and yet leaves a sufficiently indefinite line of separation to enable each paleontologist to decide for himself how nearly he will approach modern times. Care should be taken not to confuse the terms "fossil" and "petrifaction." To petrify is to turn to stone, and it is evident from what has been said that not all fossils are petrified. On the other hand, not everything which is petrified is a fossil, for to be a fossil an object must be organic in origin. So far as possible, the adjective "fossilized" should be avoided. It is at best a meaningless word, usually employed as a synonym for "petrified," with which it is not synonymous; it is therefore much misused.

Because the prime necessity for the preservation of an organism is that it be protected from decay by some covering, aquatic animals and plants are much more apt to be preserved than the inhabitants of the land. Consequently, marine invertebrates and fish are much more common as fossils than terrestrial organisms are. A land animal stands little chance of becoming a fossil unless it happens to die in a bog or by a stream, although the preservation of bones in caves under the protection of a stalagmitic or earthy cover is common.

The task of the paleontologist is to reconstruct from such materials as he finds in sedimentary rocks the animal and vegetable life of the period during which the strata were in the process of accumulation. Too often he must rely upon incomplete, distorted, and broken objects which are not easily interpreted. The solution of his difficulties can come only through comparison: first, with other materials from the same strata, in the endeavor to bring together scattered parts belonging to the same sort of animal or plant; and second, with such modern organisms as appear to be

Slab of nonmarine shale containing well-preserved fossil ferns which flourished nearly 300 million years ago. (Museum of Comparative Zoology, Harvard University.)

related. A wide knowledge of the comparative anatomy of modern organisms is therefore a necessary part of the equipment of the paleontologist. Fragmentary material must be pieced together to build up a whole skeleton. Then from scars of muscles, shapes of bones or shells, and such other features as may be preserved, it may be possible to arrive at a rough approximation of the form of the animal as it appeared in life. Such restorations, however, are always to be labeled as tentative, for the entire anatomy of few extinct animals is satisfactorily determined. Some things, such as coloration, length of fur on mammals, amount of fat, and the like, must be inferred almost entirely from our knowledge of living creatures. We cannot prove, for instance, that any fossil camel had a hump. On the other hand, some soft parts, although boneless themselves, may be recorded; for example, the proboscis of an elephant is registered by recognizable modifications of the nasal bones of the skull.

Knowledge of paleontology obviously progresses not in a single, direct line but by irregular steps along a wide frontier. New materials are constantly coming to light, revealing not only new kinds of animals and plants, never before seen by man, but new facts about extinct organisms long imperfectly known. Each so-called species of extinct creature is really an artificial creation, built up by man on the basis of remains found in the rocks. Each represents the sum total of the best information available at the time, but every scientist admits that his conception of a species is liable to change with the discovery of new material or new methods of study. It frequently happens that what is called a single species by the original describer will be seen as four, five, or ten species by some later worker with a wider knowledge of the subject.

If one is to be able to speak of any particular kind of animal or plant, it must have a name; so to each kind, or species, a name is given by the person who first publishes a description of it. Many experiments were made before a definite system of nomenclature was finally reached, about the year 1758. It was natural to try not only to assist the memory by applying a descriptive name but also to indicate the relationship of the particular animal or plant to other organisms. Men naturally like to get their knowledge into as orderly, usable, and easily remembered form as possible, and so with the naming there became involved the idea of classification. It is obvious that certain modern animals are more closely related to one animal than to another. Anyone would say at first glance that a cat and a tiger are more closely related to each other than either is to a cow. Yet the cow is more like a cat than it is like a fish. Thus the classification of plants or animals is built up about degrees of likeness or unlikeness. The name given to the particular kind is intended to furnish a knowledge of at least one degree of relationship, as well as to serve as a convenience in mentioning it. The earlier writers, who were unnecessarily descriptive, gave names a whole sentence long. Linnaeus, the great Swedish naturalist, about 1758 set the fashion now followed of cutting the name down to two words, the first or generic name indicating relationship, and the second or specific name suggesting, ideally, some outstanding characteristic of the organism described. It is usual to compare the generic name with the family name among people and the specific name with the Christian name. The generic name is given to a group of species which are found to be very closely related in the structure of their bones, teeth, muscles, etc., but each of the various kinds within a group has a specific name. Thus the genus of the cats is *Felis*; the lion is *Felis leo*, the tiger *Felis tigris*, and the house cat *Felis catus*. The generic name shows their evident close relationship, and the specific name indicates which particular kind of cat is meant.

The world at one time carried no life of any kind, but today it is teeming. How and when did life first appear? Was it different from that of today and, if so, how did it change? These are questions that have long plagued man.

Genesis, in poetic and symbolic language, tells of the appearance of life, first vegetation, then swarms of living creatures and "great sea monsters" from the waters, then birds, and finally man. Geologists have reconstructed the sequence of events and recorded it in pedantic phrases, but it is the same in broad outline. No piling up of words can conceal the miraculous majesty of the development of life.

The first living unit may well have been a viruslike form neither "plant" nor "animal." Compared to this, the amoeba, often used as an example of life at its simplest, is an extremely complex organism which has solved in its single cell the essential problems of living. From viruslike life to amoeba may have been nearly as long a step in evolution as from one-celled amoeba to many-celled man.

The following selection discusses the first living things and probable forms of life during eons that preceded what its author calls "the first and perhaps the greatest crisis in the recorded history of life, the crisis that begins the continuous record" that is documented by a fairly complete fossil record of the past half-billion years.
It is adapted from a book, **The Meaning of Evolution,** by George Gaylord Simpson, vertebrate paleontologist of The American Museum of Natural History and Columbia University.

George Gaylord Simpson

THE DEVELOPMENT OF LIFE

The first living things were almost certainly microscopic in size and not apt for any of the usual processes of fossilization. It is unlikely that any preserved trace of them will ever be found, or recognized. Indeed, it is improbable that the discovery of such remains, if any do exist, would greatly advance knowledge of how life originated. At this lowest level little could be learned from the preserved form.

However, the problem is open to study, and though the solution has not been reached it may be near, or distant, or even unattainable. Recent studies of cell structure, viruses, and gene action are converging hopefully on the mystery; they show that under conditions that may well have existed early in the history of the earth there might have been a chance organization of a complex carbon-containing molecule capable of influencing or directing the synthesis of other units like itself. Such a unit would be, in barest essentials, alive. It would be similar or analogous to a virus.

Above the level of the virus, if that be granted status as an organism, the simplest living unit is almost incredibly complex. It has become commonplace to speak of evolution from amoeba to man, as if the amoeba were a natural and simple beginning of the process. On the contrary, if life arose as a living molecule, the progression from this stage to that of the amoeba is at least as great as from amoeba to man. All the essential problems of living organisms are solved in the one-celled amoeba, and they are only elaborated in man or other multicellular animals. The step from nonlife to life may not have been so complex after all, and that from cell to multicellular organism is readily comprehensible. The change from virus to amoeba was probably the most complex that has occurred in evolution, and it may well have taken as long as the change from amoeba to man. In the billions of years of earth history the time is available even for this.

Although the rocks of 3 billion years are available for our inspection, only those found during the last sixth of this time contain reasonably satisfactory evidence of life. Some early Cambrian rocks, laid down about 500 million years ago, are crowded with fossils. In one place or another on the earth there are rich fossil deposits of almost all ages since this time. But in rocks earlier than the Cambrian, representing the great span of 2,500 million years, fossils are generally lacking and even when present are rare and usually disputed.

Aside from some vague trails, and supposed fossils with even more faulty credentials, the only definite Pre-Cambrian records are those of algae—simplest plants, but not the lowliest, for they were already a little way along the path toward higher forms. There is no serious dispute regarding their occurrence from at least the middle part of the long Pre-Cambrian span onward.

Most Pre-Cambrian rocks have been so altered as to be unsuitable for the clear preservation of fossils. This, however, is not true of all, and the exceptions have been so carefully searched that fossils other than algae should have been found if present. There must be some special reason why varied fossils are suddenly present in the Cambrian and not before.

This major mystery of the history of life has naturally excited a great deal of argument and speculation. The generally held theory is that the groups that appeared in the Cambrian really arose slowly and well back in the Pre-Cambrian, but that the earlier forms had no shells or other hard parts to be preserved.

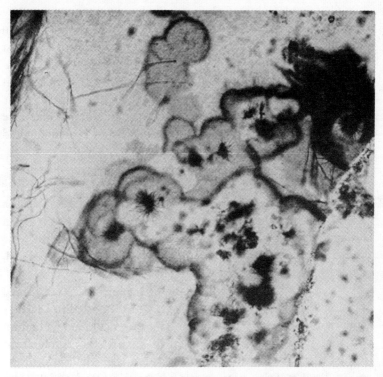

Photograph of algae from southern Ontario, Canada. These are undistorted remains of primitive plants which existed 3,500 million years ago—magnified 325 times. (S. A. Tyler and E. S. Barghoorn.)

Conditions permitting preservation of soft parts do occasionally exist—there are some remarkable examples from the Cambrian itself—but they are so exceptional that the absence of such deposits in the Pre-Cambrian would not be surprising. As to how it happened (if it did) that Pre-Cambrian animals did not have hard parts and their Cambrian descendants did, we do not know. But it is probable that development of hard parts and of other diagnostic characters of groups distinguished by these occurred as parts of a single, rather rapid evolutionary process; and that the development of hard parts by different groups at very approximately the same time was not pure coincidence.

The simultaneity of these events has been exaggerated by the too-sweeping statement that most of the animal phyla known as fossils appear in the Cambrian. The Cambrian was the longest single period of those into which geologists divide the past 5 billion years of earth history. Its length is variously estimated at from 60 million to 90 million years. Even the early part of the period had a duration of not less than 20 million years. These are long times, even to a geologist, and a great deal of evolution could occur in them, even at moderate rates.

Thirty million years ago our ancestor was something like a primitive ape and 60 million years ago something like a tree shrew. The various Cambrian animal phyla do not all appear as fossils in the very earliest rocks of that period; they straggle in throughout. As a whole, the early Cambrian representatives of the groups that did appear then are markedly simple and generalized, as if near the origin of their respective lines.

Relatively simple, small, soft-bodied animals of several basic types and myriad detailed sorts had probably existed in the Pre-Cambrian for a long time—how long we may never know. During that time the marvels of the origin of life and the development of the cell and cellular organisms had been wrought. Toward the end of Pre-Cambrian time, stages had been reached which were ripe for development. A crisis of rapid evolution, of divergence, of specialization for new or different major ways of life was at hand. Once the possibility existed, the event would tend to develop rapidly. It is one of the best attested generalities of evolution that its rate is exceptionally fast when an evolving group takes on some hitherto unexploited way of life. Also, interaction among the various groups involved would tend to intensify their divergence and accelerate its speed. Development of hard parts would permit, demand, and hasten structural change and would be selectively advantageous as defense or offense and as a means of developing new activities and invading new environments.

So, speculatively but not groundlessly, may be explained the first and perhaps the greatest crisis in the recorded history of life, the crisis that begins the continuous record. Regarding the long history of the Pre-Cambrian, the principles of evolution must have been the same then, though the organisms through which they worked were different. The paucity of earlier records is bitterly regretted, but there is no reason to fear seriously that what may be learned from the later record about the grand processes of evolution is untrue as regards unknown earlier parts of the history of life.

There are about a million different kinds of animals in the world today, and man is superior to all. Because of our intellect, we cannot help being curious about ourselves, about where we fit into the picture.

We are the latest members of a large group called vertebrates, characterized by the possession of backbones. However, an account of our development becomes meaningful only when we fit it into a framework of time so vast that it is given a special name—geologic time.

For reasons still in debate, fossils began to accumulate in large numbers only 500 million years ago at the start of a geologic period called the Cambrian. At this time, almost all invertebrate groups were already developed, and animals were highly organized. Unfortunately, fossils of the immediate predecessors of the first vertebrates have not been found for us to study, but there is living today a sea lancelet known as Amphioxus that we believe is much like this ancestor. The first vertebrates appeared and lived in the sea 400 million years ago; dinosaurs started 200 million years ago, ruled the earth for a time, and disappeared 70 million years ago; mammals took over, and 1 million years ago earliest man appeared. "In the last 100,000 years he has gone far. . . ." Yet in true perspective, man took up his assignment in the scheme of things only yesterday.

Edwin H. Colbert, Chairman of the Department of Geology and Paleontology of The American Museum of Natural History, tells this story in the following selection adapted from his book **Evolution of the Vertebrates.**

Edwin H. Colbert

MAN AND GEOLOGIC TIME

We shall probably never know what the first vertebrates were like, because it is unlikely that adequate indications of them are preserved in the fossil record. They must have been small, comparatively simple animals, and it is not likely that they had a hard skeleton, capable of being fossilized and preserved in the sediments of the earth. They first appear in the record of the rocks with highly developed bony armor, and for this reason it has been argued that bone is very primitive in the history of vertebrate evolution. Yet it seems logical to believe that there might have been a long period of vertebrate evolution preceding the development of bony armor so that the first ancient, bone-encrusted vertebrates in the geologic record in truth may be advanced far beyond the condition of their primitive ancestors.

It happens, however, that there is a modern animal of such primitive form and organization that it approximates to a considerable degree our conception of the central ancestor for the backboned animals. This is the little sea lancelet known as *Amphioxus*, which lives in the shallow waters along certain tropical coasts, and spends much of its life buried in the sandy bottom. It is a translucent animal of fishlike form and rarely more than 2 inches in length. There are no vertebrae in *Amphioxus*, but rather a continuous chord forming an internal support for the animal.

This support represents the precursor of the backbone or spinal column. Above this elastic rod is the nerve chord; below it is a simple digestive tube. There is no real head or brain in this little animal, and no sense organs, except for some pigment spots that seem to be sensitive to light. No matter how primitive and deficient *Amphioxus* is in these respects, it is well supplied with gills, arranged in a long series down either side of the front portion of the body. These extended gills serve to extract oxygen from the water; they also aid the animal in feeding, by functioning as a sort of sieve to strain food from the debris of the ocean floor. Finally, although it is a capable swimmer, it lacks paired fins of any sort, except for a small tail fin.

The great development of the gill basket shows that this animal may be specialized in some aspects of its anatomy, yet in spite of this specialization the lancelet remains in a general way an approximate structural ancestor for the vertebrates. It is probable that there has been little change in the line of evolution represented by the lancelet since early Paleozoic or even Pre-Paleozoic times. In *Amphioxus* we see in effect our ancestor of 500 million years ago.

Mention of the word Paleozoic and reference to a time span of such tremendous duration as 500 million years brings us to the subject of geologic time. This is a consideration of prime importance to the paleontologist, because one of the great advantages that the study of fossils has over other branches of natural history is the possibility of projecting ourselves back through the fourth dimension of time into the past ages of the earth.

It is not easy at first to think in the immense units of geologic time. We are used to thinking in terms of years or centuries or millenniums; geologic time is measured in millions of years. It seems almost incredible that there were rains and winds and volcanoes and the cycles of life and death on the earth as far back as 100 million years or 500 million years or a billion years ago, yet from the study of radioactive elements we know that such great time spans are necessary to measure the sequence of events that make up earth history.

It is not possible at this place to go into the evidence for dating the earth or for drawing up the geologic time scale, but a few salient facts can be presented. As already mentioned, the first rocks that can be dated are almost 3 billion years old. Fossils first appear in abundance in rocks of Cambrian age, and they can be dated as about 500 million years old. Although this is the beginning of an adequate fossil record, it is by no means any indication of the beginning of life on the earth, for there was probably a long time span before the Cambrian period when plants and animals were evolving as very primitive organisms. By the beginning of Cambrian times life was sufficiently advanced so that animals were highly organized, with hard parts capable of fossilization. Almost all the major groups of invertebrate animals are present in Cambrian rocks—an indication of the incredibly long evolutionary sequence during which life was differentiating and specializing to the comparatively high degree that is characteristic of the earliest Cambrian faunas.

Perhaps a few other dates will help to indicate the long evolutionary history of life on the earth. The first vertebrates appeared in earth history some 400 million years ago. The dinosaurs began their long evolutionary history almost 200 million years ago, and they continued for more than 100 million years. They became extinct about 70 million years ago, and at that time mammals became the dominant animals on the earth. Man as such appeared less than a million years ago. In the last few hundred thousand years he has gone far.

Several eras mark the course of earth history prior to the beginning of the fossil records, but since this portion of geologic time is rather difficult to interpret, the general practice is to refer to it as Pre-Cambrian times. The Pre-Cambrian portion of earth history is long, extending through more than 2,500 million years of time.

With the beginning of the fossil record, earth history can be measured and followed in considerable detail. As a result of cumulative studies carried on during the last century or so, three great eras of earth history are recognized by the sequence of the fossil

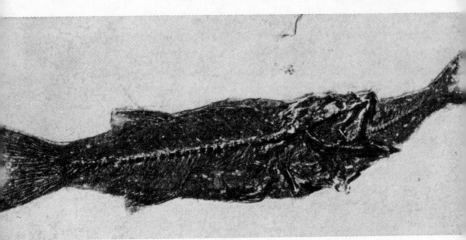

This specimen from Fossil, Wyoming, records an episode in earth history that occurred about 60 million years ago. Involved in this ancient error are a predatory perch and its intended victim, a herring. They lived in ancient Lake Gosiute, a large, shallow, fresh-water body that occupied the region of present Green River Basin in southwestern Wyoming and adjacent areas. (Fred Anderegg.)

record. They are, in the order of their age, the Paleozoic era, the time of ancient life; the Mesozoic era, the period of middle life; and the Cenozoic era, the time of recent life.

As mentioned above, the first vertebrates known are from rocks of early Paleozoic age. However, it is not until we reach the sediments deposited during mid-Paleozoic that fossils are complete enough to give us some idea as to the form and relationships of the early vertebrates. From then on, however, their history is well known, as it is revealed in successively younger layers of the earth's crust. All the major groups of fishes had appeared by this time, some of them to continue to the present day and one group to become extinct before the close of the Paleozoic era. The first land-living vertebrates, the amphibians, made their appearance as descendants of certain advanced fishes. The amphibians had their heyday, especially during the final stages of the Paleozoic era, after which they continued to develop, but on a smaller scale.

In the course of their evolution the amphibians gave rise to

GEOLOGIC CALENDAR

Eras	Eras began, years ago	Periods	Evolution of life		
			Animals		Plants
Cenozoic	60 million	Pleistocene Pliocene Miocene Oligocene Eocene Paleocene	Man Horses		Grasses become abundant Flowering land plants
Mesozoic	200 million	Cretaceous Jurassic Triassic	Extinction of dinosaurs Birds first appear Dinosaurs first appear		Cone-bearing land plants
Paleozoic	500 million	Permian Pennsylvanian Mississippian Devonian Silurian Ordovician Cambrian	Amphibians Fishes (first vertebrates) Marine invertebrates (first abundant fossil record)		Spore-bearing land plants
Pre-Cambrian	(Oldest rocks now known were formed.) 3,000 million	Rocks abundant but world-wide names not firmly established	Primitive invertebrates (scanty fossil record) One-celled organisms		Marine plants Algae

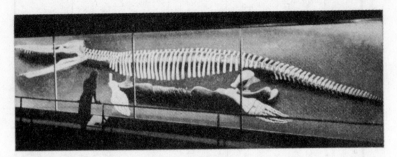

The sea monster Kronosaurus, a reptile which flourished in the seas more than 100 million years ago, when dinosaurs were roaming the lands. Its bones were found in Australia in 1932 and shipped to Harvard University where the painstaking job of chipping away the rock in which the bones were embedded was carried out over the next 25 years. (Harvard University News Office.)

the reptiles, which were destined to rule the earth for many millions of years. During the Mesozoic era some reptiles, known as dinosaurs, became the dominant animals on the land, evolving along diverse lines that carried them to all the larger land areas of the earth and into most of the different environments that then existed. The dinosaurs were indeed the rulers of the earth for more than 100 million years, but finally they became extinct. Today the reptiles that survive, although numerous and widely dispersed throughout the world, are but a remnant of the hordes that once ruled the earth.

Before the dinosaurs had become extinct, in fact during the earlier part of their long evolutionary history, two other groups of vertebrates arose from reptilian ancestors. They evolved, slowly at first and then with increasing rapidity and diversity, into the birds and the mammals of our present-day world. By late Mesozoic, the birds had become highly specialized; by the beginning of the Cenozoic era they were populating the continents and the islands of the world in essentially their modern form. As long as the reptiles were dominant the evolution of the mammals was comparatively slow. But with the transition from the Mesozoic era into the Cenozoic era new opportunities were opened to the early mammals. The reign of the reptiles was at an end, and the

Age of Mammals began. It has continued to this day, when one mammal, man, has developed within a comparatively short span of time to unprecedented heights. This is now the age of intellect (no matter how despairing some people may feel about modern world trends), and because of his intellect the mammal known as man is able to look back through time to study the evolutionary history of his forerunners and forebears.

Our world of geology would mean nothing to us without the atmosphere, which is indispensable to the maintenance of life. The atmosphere is also an important geologic agent as it contributes to the shaping of the land. But we wonder whether it has always been the same, where its oxygen came from, why it does not stream off into outer space. Actually, the first atmosphere did escape into space, but it was good riddance as it consisted of compounds that could neither develop nor support life. Then as the earth cooled and finally solidified, its molten materials gave off gases that had been dissolved in them, and a new atmosphere, mostly of water vapor, carbon dioxide, and nitrogen, accumulated and remained. There was no oxygen. It was during this phase that plant life developed with its unique capability of converting carbon dioxide into oxygen. This aspect of plant life made possible and eventually led to the development of animal life and the formation of our present atmosphere, which is still changing today.

The story of the origin and evolution of the atmosphere is told in the following selection adapted from the article "The Origin of the Atmosphere," by Dr. Helmut E. Landsberg, which was published in **Scientific American.** Dr. Landsberg, a geophysicist, founded the Geophysical and Meteorological Laboratory of Pennsylvania State College and was later Director of Geophysics Research at the Air Force Cambridge Research Center.

Helmut E. Landsberg

THE ORIGIN OF THE ATMOSPHERE

Most of us take the earth's air for granted and, except for slight annoyance at some weather, regard it as stable. Yet the atmosphere has undergone dramatic evolution and it is still changing slowly.

In the beginning, when the earth's surface was much hotter than it is today, the gases surrounding it were compounds of chlorine, fluorine, bromine, iodine, and sulfur, as well as ammonia. These had their molecules in such rapid motion because of the high temperature (14,500°F) that they quickly escaped into space.

As the earth cooled, and finally solidified, the molten material gave off gases which had been dissolved in it. Among these were water vapor, carbon dioxide, and nitrogen, which comprised at least 90 per cent of the atmosphere at that time. No free atoms of oxygen were present during this stage of development; but now free oxygen is the second most abundant element in our atmosphere and the one, of course, most important to life. Where did it come from? The most likely explanation is that it has been accumulated and maintained by the photosynthesis reaction of carbon dioxide in plants.

With further cooling, the atmosphere reached the critical

temperature (705°F) at which liquid water can exist simultaneously with water vapor. Then, as cooling continued, oceans formed. Once the temperature got below the boiling point of water (212°F) most of the water vapor condensed. At this stage, the composition was 74 per cent carbon dioxide, 15 per cent water vapor, and 10 per cent nitrogen.

From here on, changes in the atmosphere became very slow indeed. There was gradual feeding in of gases from volcanic processes, along with interactions of weather processes on land, and solution and sedimentation processes in the seas.

When a temperature of 160°F was reached, life processes began. We cannot cover all the speculations on this point, but it is important to note that some cells can live in an atmosphere without oxygen. The present purple sulfur bacteria are an example of this, producing organic matter in a carbon dioxide atmosphere. Some such organisms may have been the first living things on earth, preparing the way for the earliest green plants, which in turn would supply most of the free oxygen to support higher forms of life. Further evolution of plants would remove about 40 per cent of the carbon dioxide still in the air, while the rest went into the deposition of carbonate rocks. Perhaps 2 billion years were required to transform the atmosphere into the present nitrogen-oxygen mixture. During the past billion years, this atmosphere has been essentially in a state of equilibrium, with 99 per cent of the air mass below 25 miles, though traces of air have been observed well above 600 miles. Production and consumption of the various gases balance. The major producer in the process is volcanic action; the stabilizers are plants and the oceans.

At the earth's surface today, the air we breathe is a mixture of 21 per cent oxygen, 78 per cent nitrogen, 0.033 per cent carbon dioxide, and the rare gases argon, neon, krypton, and xenon. This mixture (up to 45 miles) is nearly constant in proportions, with only minute local and seasonal fluctuations. Other constituents of the lower atmosphere, especially water vapor and dust particles, are quite variable, changing rapidly with altitude. Above 6 to 9 miles, there is good evidence that water vapor is almost com-

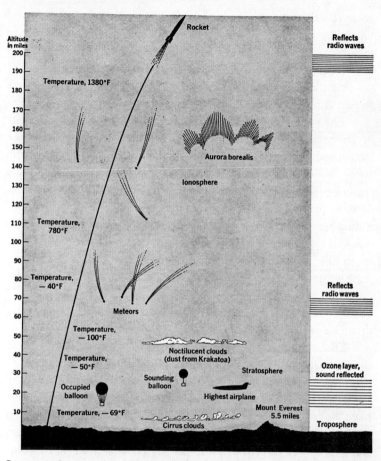

Some of the aspects of the first 200 miles of our present atmosphere. (After J. C. Hogg, J. B. Cross, and K. Vordenberg, Physical Science, D. Van Nostrand Company, Inc., New York, 1959.)

pletely absent, and dust decreases rapidly the higher we go. At about 14 miles, there is a layer of ozone which absorbs much of the burning ultraviolet radiation from the sun; any sudden removal of this layer could threaten all life on earth.

At the outer limits of the atmosphere, molecules and atoms become electrically unbalanced due to the bombardment of solar and cosmic radiations. During the hours of darkness, many of these reactions reverse themselves, causing a continual stirring and changing of physical form. These ionospheric phenomena are mainly of interest in studying radio propagation, but they also have an important bearing on the first problem to be met in considering the origin of the atmosphere: Why hasn't the atmosphere streamed off into outer space?

The escape of gases from the earth is governed by two sets of factors. One is their temperature and density; the other is the size, mass, and gravitational pull of the earth. To overcome the latter, a body shot upward must exceed a speed of 7 miles a second to escape into space.[1] The average velocity of molecules of nitrogen and oxygen in the earth's atmosphere is a quarter of a mile per second. The speed and direction of individual molecules are governed by the frequency of collision with other molecules, which in turn is determined by heat and density. Thus, it becomes a statistical problem as to how often individual molecules in collision near the boundary of the atmosphere will achieve velocities sufficiently higher than average, in the right direction, to escape from the earth.

From this, we may see that the earth can, indeed, hold its present atmosphere for a long time. Nothing in this implies, however, that there will be no further changes in its composition, and we may well ask where we are going from here. Some scientists have looked at the other planets to find an answer, reasoning that one or another of them may have passed the present evolutionary stage of the earth. Most of the planets on which we have information, such as Venus, Jupiter, and Saturn, seem to have atmospheres still in a much earlier phase than the earth's. The only one that appears to fit a later stage is Mars, but there are major initial differences between Mars and the earth. One is its lower density, hence much lower escape velocity. The present Martian atmosphere seems to be mainly nitrogen, argon, carbon dioxide, and water vapor. No oxygen has been noted in the spectrograms.

---

[1] Escape velocities for other planets range from only 1.5 miles per second on the moon, 3.1 on Mars, and 6.5 on Venus, to 38 miles per second on giant Jupiter.

Whatever oxygen may have been present has either escaped or has been used up in the oxidation of rocks. But it is not probable that the earth's oxygen will escape and it is a moot question whether it will be used up, so Mars provides only the cloudiest of answers.

How does human activity influence the earth's atmosphere? There is some evidence that industrial life has increased the amount of carbon dioxide in the air, while in certain areas such gases as nitrous oxide, carbon monoxide, and sulfur dioxide are definitely on the increase. Solid suspensions and radioactive debris are also becoming more prominent. So far, however, nature holds the upper hand and cleanses many of the pollution products from the atmosphere with rainfall. Meanwhile, photosynthesis still adds oxygen to the air in a mighty stream. It is only reasonable to conclude that further billions of years may elapse before another atmosphere, with its possible composition of nitrogen, argon, carbon dioxide, and water vapor, spells a grand-scale *finis* to most terrestrial life.

Oceans not only mothered the first life, the only life for eons, but in them were once gathered the sedimentary materials now found blanketing three-fourths of the land surfaces and towering in the highest mountains. So they are in a real sense the essence of an important part of our world of geology.

It was not until the sixteenth century that a constructive effort toward solving some of the ocean's mysteries was begun. This was when Magellan tried unsuccessfully to sound the depths of the Pacific Ocean with a 1,200-foot line. After that, knowledge of the oceans developed, but slowly; in the present century it spurted forward as interest grew and a wealth of information was acquired. We have learned that the average depth of the oceans is 12,400 feet, in contrast to the land's average elevation of 2,700 feet; and the oceans cover more than 70 per cent of the globe. There is a lot of water out there concealing secrets that geologists continually seek to ferret out.

In some areas, we have determined the shape of the ocean floor by bouncing sound waves from it, mapped currents and tides, studied abundant sea life, determined higher and lower past stands of the sea's surface. We have even gone so far as to assume from the earth's history that 5 billion years or so ago there were no oceans, as the surface cooled from a bubbling expanse of molten rock enveloped by noxious gases to a solid surrounded by water vapor, carbon dioxide, and nitrogen but so hot that it turned water to steam on contact. As the rocks cooled further, rains fell, water accumulated, and the oceans had their beginning. The water was fresh at the start, but as it coursed over rocks on the way to gathering places it dissolved minerals and carried them to the growing seas.

Water is in constant circulation. It falls from the clouds, moves over the land, enters the ocean, and is once again evaporated to form new clouds. In the ocean, some of its movements are governed by the pull of the moon, by winds, and by temperature differences. The moon influences the tides, which range from nearly 40 feet in the Bay of Fundy to almost imperceptible rises in many other places. Winds rumple the surface into combing waves, and trade winds drive currents along the surface like rivers in the sea. Other currents result when cold water at the poles sinks to the bottom and drifts toward the equator.

Saltiness of the sea varies. It is least in regions of heavy rainfall or melting glaciers and highest where evaporation is rapid and rainfall slight. These and other features of the ocean story have never been better told than by Rachel Carson in her book **The Sea Around Us,** which has been described as "unique, both for its authentic comprehensive information, and for its imaginative, poetic writing." The following account is selected from that classic.

Rachel L. Carson

OCEANS AND THEIR HIDDEN LANDS

The first European ever to sail across the wide Pacific was curious about the hidden worlds beneath his ship. Between the two coral islands of St. Paul and Los Tiburones in the Tuamotu Archipelago, Magellan ordered his sounding line to be lowered. It was the conventional line used by explorers of the day, no more than 200 fathoms long. It did not touch bottom, and Magellan declared that he was over the deepest part of the ocean. Of course he was completely mistaken, but the occasion was none the less historic. It was the first time in the history of the world that a navigator had attempted to sound the depths of the open ocean.

Three centuries later, in the year 1839, Sir James Clark Ross set out from England in command of two ships with the names of dark foreboding, the *Erebus* and the *Terror*, bound for the "utmost navigable limits of the Antarctic Ocean." As he proceeded on his course, he tried repeatedly to obtain soundings, but failed for lack of a proper line. Finally he had one constructed on board, of "three thousand six hundred fathoms, or rather more than four miles in length. . . . On the 3rd of January, in latitude 27° 26′ S., longitude 17° 29′ W., the weather and all other circumstances being propitious, we succeeded in obtaining soundings with two thousand four hundred and twenty-five fathoms of line, a depression of the bed of the ocean beneath its surface very little short of the elevation of Mont Blanc above it." This was the first successful abyssal sounding.

But taking soundings in the deep ocean was, and long remained, a laborious and time-consuming task, and knowledge of the undersea topography lagged considerably behind our acquaintance with the landscape of the near side of the moon. Over the years, methods were improved. For the heavy hemp line used by Ross, Maury of the United States Navy substituted a strong twine, and in 1870 Lord Kelvin used piano wire. Even with improved gear a deep-water sounding required several hours or sometimes an entire day. By 1854, when Maury collected all available records, only 180 deep soundings were available from the Atlantic, and by the time that modern echo sounding was developed, the total that had been taken from all the ocean basins of the world was only about 15,000. This is roughly one sounding for an area of 6,000 square miles.

Now hundreds of vessels are equipped with sonic sounding instruments that trace a continuous profile of the bottom beneath

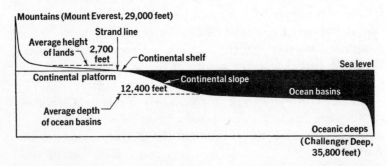

The relative height of continental areas and depth of ocean basins on a scale where the width of the diagram represents 100 per cent of the earth's surface. This shows the small fraction of the surface covered by mountains and ocean deeps, as well as the broad distinction between continental platform and ocean-basin areas. (After Raymond C. Moore, Introduction to Historical Geology, *2d ed., McGraw-Hill Book Company, Inc., New York, 1958.)*

the moving ship (although only a few can obtain profiles at depths greater than 2,000 fathoms). Soundings are accumulating much faster than they can be plotted on the charts. Little by little, like the details of a huge map being filled in by an artist, the hidden contours of the ocean are emerging. But, even with this recent progress, it will be years before an accurate and detailed relief map of the ocean basins can be constructed.

The general bottom topography is, however, well established. Once we have passed the tide lines, the three great geographic provinces of ocean are the *continental shelves*, the *continental slopes*, and the *floor of the deep sea*. Each of these regions is as different from the other as an arctic tundra from a range of the Rocky Mountains.

The *continental shelf* is of the sea, yet of all regions of the ocean it is most like the land. Sunlight penetrates to all but its deepest parts. Plants drift in the waters above it; seaweeds cling to its rocks and sway to the passage of the waves. Familiar fishes —unlike the weird monsters of the abyss—move over its plains like herds of cattle. Much of its substance is derived from the land—the sand and the rock fragments and the rich topsoil carried by running water to the sea and gently deposited on the shelf. Its submerged valleys and hills, in appropriate parts of the world, have been carved by glaciers into a topography much like the northern landscapes we know and the terrain is strewn with rocks and gravel deposited by the moving ice sheets. Indeed many parts (or perhaps all) of the shelf have been dry land in the geologic past, for a comparatively slight fall of sea level has sufficed, time and again, to expose it to wind and sun and rain. The Grand Banks of Newfoundland rose above the ancient seas and were submerged again. The Dogger Bank of the North Sea shelf was once a forested land inhabited by prehistoric beasts; now its "forests" are seaweeds and its "beasts" are fishes.

Of all parts of the sea, the continental shelves are perhaps most directly important to man as a source of material things. The great fisheries of the world, with only a few exceptions, are confined to the relatively shallow waters over the continental shelves. Seaweeds are gathered from their submerged plains to make scores of substances used in foods, drugs, and articles of commerce. As the petroleum reserves left on continental areas by ancient seas become depleted, petroleum geologists look more and more for the oil that may lie, as yet unmapped and unexploited, under these bordering lands of the sea.

The shelves begin at the tidelines and extend seaward as gently sloping plains. The 100-fathom contour used to be taken as the boundary between the continental shelf and the slope; now it is customary to place the division wherever the gentle declivity of the shelf changes abruptly to a steeper descent toward abyssal depths. The world over, the average depth at which this change occurs is about 72 fathoms; the greatest depth of any shelf is probably 200 to 300 fathoms.

Nowhere off the Pacific coast of the United States is the continental shelf much more than 20 miles wide—a narrowness characteristic of coasts bordered by young mountains perhaps still in the process of formation. On the American east coast, however, north of Cape Hatteras the shelf is as much as 150 miles wide. But at Hatteras and off southern Florida it is merely the narrowest of thresholds to the sea. Here its scant development seems to be related to the press of that great and rapidly flowing river-in-the-sea, the Gulf Stream, which at these places swings close inshore.

The widest shelves in all the world are those bordering the Arctic. The Barents Sea shelf is 750 miles across. It is also relatively deep, lying for the most part 100 to 200 fathoms below the surface, as though its floor had sagged and been downwarped under the load of glacial ice. It is scored by deep troughs between which banks and islands rise—further evidence of the work of the ice. The deepest shelves surround the Antarctic continent, where soundings in many areas show depths of several hundred fathoms near the coast and continuing out across the shelf.

Once beyond the edge of the shelf, as we visualize the steeper declivities of the *continental slope*, we begin to feel the mystery and the alien quality of the deep sea—the gathering darkness, the growing pressure, the starkness of a seascape in which all plant life has been left behind and there are only the unrelieved contours of rock and clay, of mud and sand.

Biologically the world of the continental slope, like that of the abyss, is a world of animals—a world of carnivores where each creature preys upon another. For no plants live here, and the only ones that drift down from above are the dead husks of the flora of the sunlit waters. Most of the slopes are below the zone of surface wave action, yet the moving water masses of the ocean currents press against them in their coastwise passage; the pulse of the tide beats against them; they feel the surge of the deep, internal waves.

Geographically, the slopes are the most imposing features of all the surface of the earth. They are the walls of the deep-sea basins. They are the farthermost bounds of the continents, the true place of beginning of the sea. The slopes are the longest and highest escarpments found anywhere on the earth; their average height is 12,000 feet, but in some places they reach the immense height of 30,000 feet. No continental mountain range has so great difference of elevation between its foothills and its peaks.

Nor is the grandeur of slope topography confined to steepness and height. The slopes are the site of one of the most mysterious features of the sea. These are the submarine canyons with their steep cliffs and winding valleys cutting back into the walls of the continents. The canyons have now been found in so many parts of the world that when soundings have been taken in presently unexplored areas we shall probably find that they are of world-wide occurrence. Geologists say that some of the canyons were formed well within the most recent division of geologic time, the Cenozoic, most of them probably within the Pleistocene, a million years ago, or less. But how and by what they were carved, no one can say. Their origin is one of the most hotly disputed problems of the ocean.

Only the fact that the canyons are deeply hidden in the darkness of the sea (many extending a mile or more below present sea level) prevents them from being classed with the world's most spectacular scenery. The comparison with the Grand Canyon of the Colorado is irresistible. Like river-cut land canyons, sea canyons are deep and winding valleys, V-shaped in cross section, their walls sloping down at a steep angle to a narrow floor. The location of many of the largest ones suggests a past connection with some of the great rivers of the earth of our time. Hudson Canyon, one of the largest on the Atlantic coast, is separated by only a shallow sill from a long valley that wanders for more than a hundred miles across the continental shelf, originating at the entrance of New York Harbor and the estuary of the Hudson River. There are large canyons off the Congo, the Indus, the Ganges, the Columbia, the São Francisco, and the Mississippi, according to Francis Shepard, one of the principal students of the canyon problem. Monterey Canyon in California, Professor Shepard points out, is located off an old mouth of the Salinas River; the Cap Breton canyon in France appears to have no relation to an existing river, but actually lies off an old fifteenth-century mouth of the Adour River.

Their shape and apparent relation to existing rivers have led Shepard to suggest that the submarine canyons were cut by rivers at some time when their gorges were above sea level. The relative youth of the canyons seems to relate them to some happenings in the world of the Ice Age. It is generally agreed that sea level was lowered during the existence of the great glaciers, for water was withdrawn from the sea and frozen in the ice sheets. But most geologists say that the sea was lowered only a few hundred feet—not the mile that would be necessary to account for the canyons. According to one theory, there were heavy submarine mud flows during the times when the glaciers were advancing and sea level fell the lowest; mud stirred up by waves poured down the continental slopes and scoured out the canyons. Since none of the present evidence is conclusive, however, we simply do not know how the canyons came into being, and their mystery remains.

The floor of the deep ocean basins is probably as old as the sea itself. In all the hundreds of millions of years that have intervened since the formation of the abyss, these deeper depressions have never, as far as we can learn, been drained of their covering waters. While the bordering shelves of the continents have known, in alternating geologic ages, now the surge of waves and again the eroding tools of rain and wind and frost, always the abyss has lain under the all-enveloping cover of miles-deep water.

But this does not mean that the contours of the abyss have remained unchanged since the day of its creation. The floor of the sea, like the stuff of the continents, is a thin crust over the [plastic mantle] of the earth. It is here thrust up into folds and wrinkles as the interior cools by imperceptible degrees and shrinks away from its covering layers; there it falls away into deep trenches in answer to the stresses and strains of crustal adjustment; and again it pushes up into the conelike shapes of undersea mountains as volcanoes boil upward from fissures in the crust.

Until very recent years, it has been the fashion of geographers and oceanographers to speak of the floor of the deep sea as a vast and comparatively level plain. The existence of certain topographic features was recognized, as, for example, the Atlantic Ridge and a number of the very deep depressions like the Mindanao Trench off the Philippines. But these were considered to be rather exceptional interruptions of a flat floor that otherwise showed little relief.

This legend of the flatness of the ocean floor was thoroughly destroyed by the Swedish Deep-Sea Expedition, which sailed from Göteborg in the summer of 1947 and spent the following 15 months exploring the bed of the ocean. While the *Albatross* of this expedition was crossing the Atlantic in the direction of the Panama Canal, the scientists aboard were astonished by the extreme ruggedness of the ocean floor. Rarely did their fathometers reveal more than a few consecutive miles of level plain. Instead the bottom profile rose and fell in curious steps constructed on a Gargantuan scale, half a mile to several miles wide. In the Pacific, the uneven bottom contours made it difficult to use many of the oceanographic instruments. More than one coring tube was left behind, probably lodged in some undersea crevasse.

One of the exceptions to a hilly or mountainous floor was in the Indian Ocean, where, southeast of Ceylon, the *Albatross* ran for several hundred miles across a level plain. Attempts to take bottom samples from this plain had little success, for the corers were broken repeatedly, suggesting that the bottom was hardened lava and that the whole vast plateau might have been formed by the outpourings of submarine volcanoes on a stupendous scale. Perhaps this lava plain under the Indian Ocean is an undersea counterpart of the great basaltic plateau in the eastern part of the State of Washington, or of the Deccan plateau of India, built of basaltic rock 10,000 feet thick.

In parts of the Atlantic basin the Woods Hole Oceanographic Institution's vessel *Atlantis* has found a flat plain occupying much of the ocean basin from Bermuda to the Atlantic Ridge and also to the east of the Ridge. Only a series of knolls, probably of volcanic origin, interrupts the even contours of the plains. These particular regions are so flat that it seems they must have remained largely undisturbed, receiving deposits of sediments over an immense period of time.

The deepest depressions on the floor of the sea occur not in the centers of the oceanic basins as might be expected, but near the continents. One of the deepest trenches of all, the Mindanao, lies east of the Philippines and is an awesome pit in the sea, six and a half miles deep. The Tuscarora Trench east of Japan, nearly as deep, is one of a series of long, narrow trenches that border the convex outer rim of a chain of islands including the Bonins, the Marianas, and the Palaus. On the seaward side of the Aleutian Islands is another group of trenches. The greatest deeps of the Atlantic lie adjacent to the islands of the West Indies, and also below Cape Horn, where other curving chains of islands go out like stepping stones into the Southern Ocean. And again in the Indian Ocean the curving island arcs of the East Indies have their accompanying deeps.

Always there is this association of island arcs and deep trenches, and always the two occur only in areas of volcanic unrest. The pattern, it is now agreed, is associated with mountain making and the sharp adjustments of the sea floor that accompany it. On the concave side of the island arcs are rows of volcanoes. On the convex side there is a sharp down-bending of the ocean floor, which results in the deep trenches with their broad V-shape. The two forces seem to be in a kind of uneasy balance: the upward folding of the earth's crust to form mountains, and the thrusting down of the crust of the sea floor into the basaltic substance of the underlying layer. Sometimes, it seems, the down-thrust mass of granite has shattered and risen again to form islands. Such is the supposed origin of Barbados in the West Indies and of Timor in the East Indies. Both have deep-sea deposits, as though they had once been part of the sea floor. Yet this must be exceptional. In the words of the great geologist, Daly,

> Another property of the earth is its ability . . . to resist shearing pressures indefinitely. . . . The continents, overlooking the sea bottom, stubbornly refuse to creep thither. The rock under the Pacific is strong enough to bear, with no known time limit, the huge stresses involved by the down-thrust of the crust at the Tonga Deep, and by the erection of the 10,000-meter dome of lavas and other volcanic products represented in the island of Hawaii.[1]

The least-known region of the ocean floor lies under the Arctic Sea. The physical difficulties of sounding here are enormous. A permanent sheet of ice, as much as fifteen feet thick, covers the whole central basin and is impenetrable to ships. Peary took several soundings in the course of his dash to the Pole by dog team in 1909. On one attempt, a few miles from the Pole, the wire broke with 1,500 fathoms out. In 1927, Sir Hubert Wilkins landed his plane on the ice 550 miles north of Point Barrow and obtained a single echo sounding of 2,975 fathoms, the deepest ever recorded from the Arctic Sea. Vessels deliberately frozen into the ice (such as the Norwegian *Fram* and

---

[1] R. A. Daly, *The Changing World of the Ice Age*, Yale University Press, New Haven, Conn., p. 116, 1934.

the Russian *Sedov* and *Sadko*) in order to drift with it across the basin have obtained most of the depth records available for the central parts. In 1937 and 1938 Russian scientists were landed near the Pole and supplied by plane while they lived on the ice, drifting with it. These men took nearly a score of deep soundings. . . .

By the middle 1940's, the total of soundings for deep arctic areas by all methods was only about 150, leaving most of the top of the world an unsounded sea whose contours could only be guessed. One interesting speculation to be tested by future soundings is that the mountain chain that bisects the Atlantic and has been supposed to reach its northern terminus at Iceland, may actually continue across the arctic basin to the coast of Russia. The belt of earthquake epicenters that follows the Atlantic Ridge seems to extend across the Arctic Sea, and where there are submarine earthquakes it is at least reasonable to guess that there may be mountainous topography.

A new feature on recent maps of undersea relief—something never included before the 1940's—is a group of about 160 curious, flat-topped sea mounts between Hawaii and the Marianas. It happened that a Princeton University geologist, H. H. Hess, was in command of the U.S.S. *Cape Johnson* during two years of the wartime cruising of that vessel in the Pacific. Hess was immediately struck by the number of these undersea mountains that appeared on the fathograms of the vessel. Time after time, as the moving pen of the fathometer traced the depth contours it would abruptly begin to rise in an outline of a steep-sided sea mount, standing solitarily on the bed of the sea. How they acquired their flat-topped contours is a mystery perhaps as great as that of the submarine canyons.

Unlike the scattered sea mounts, the long ranges of undersea mountains have been marked on the charts for a good many years. The Atlantic Ridge was discovered a century ago. The early surveys for the route of the trans-Atlantic cable gave the first hint of its existence. The German oceanographic vessel *Meteor*, which crossed and recrossed the Atlantic during the

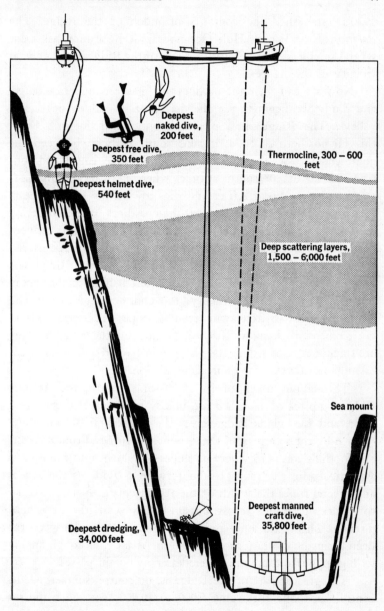

Some methods of obtaining information about the bottom of the sea. (After Stanford Research Institute Journal, Third Quarter, 1959.)

1920's, established the contours of much of the Ridge. The *Atlantis* of the Woods Hole Oceanographic Institution has spent several summers in an exhaustive study of the Ridge in the general vicinity of the Azores.

Now we can trace the outlines of this great mountain range, and dimly we begin to see the details of its hidden peaks and valleys. The Ridge rises in mid-Atlantic near Iceland. From this far-northern latitude, it runs south midway between the continents, crosses the equator into the South Atlantic, and continues to about 50° south latitude, where it turns sharply eastward under the tip of Africa and runs toward the Indian Ocean. Its general course closely parallels the coastlines of the bordering continents, even to the definite flexure at the equator between the hump of Brazil and the eastward-curving coast of Africa. To some people this curvature has suggested that the Ridge was once part of a great continental mass, left behind in mid-ocean when, according to one theory, the continents of North and South America drifted away from Europe and Africa. However, recent work shows that on the floor of the Atlantic there are thick masses of sediments which must have required hundreds of millions of years for their accumulation.

Throughout much of its 10,000-mile length, the Atlantic Ridge is a place of disturbed and uneasy movements of the ocean floor, and the whole Ridge gives the impression of something formed by the interplay of great, opposing forces. From its western foothills across to where its slopes roll down into the eastern Atlantic basin, the range is about twice as wide as the Andes and several times the width of the Appalachians. Near the equator, a deep gash cuts across it from east to west—the Romanche Trench. This is the only point of communication between the deep basins of the eastern and western Atlantic, although among its higher peaks there are other, lesser mountain passes.

The greater part of the Ridge is, of course, submerged. Its central backbone rises some 5,000 to 10,000 feet above the sea floor, but another mile of water lies above most of its summits. Yet here and there a peak thrusts itself up out of the darkness of deep water and pushes above the surface of the ocean. These are the islands of the mid-Atlantic. The highest peak of the Ridge is Pico Island of the Azores. It rises 27,000 feet above the ocean floor, with only its upper 7,000 to 8,000 feet emergent. The sharpest peaks of the Ridge are the cluster of islets known as the Rocks of St. Paul, near the equator. The entire cluster of half a dozen islets is not more than a quarter of a mile across, and their rocky slopes drop off at so sheer an angle that water more than half a mile deep lies only a few feet off shore. The sultry volcanic bulk of Ascension is another peak of the Atlantic Ridge; so are Tristan da Cunha, Gough, and Bouvet.

But most of the Ridge lies forever hidden from human eye. Its contours have been made out only indirectly by the marvelous probings of sound waves; bits of its substance have been brought up to us by corers and dredges; and some details of its landscape have been photographed with deep-sea cameras. With these aids, our imaginations can picture the grandeur of the undersea mountains, with their sheer cliffs and rocky terraces, their deep valleys and towering peaks. If we are to compare the ocean's mountains with anything on the continents, we must think of terrestrial mountains far above the timber line, with their silent snow-filled valleys and their naked rocks swept by the winds. For the sea has an inverted "timber line" or plant line, below which no vegetation can grow. The slopes of the undersea mountains are far beyond the reach of the sun's rays, and there are only the bare rocks, and, in the valleys, the deep drifts of sediments that have been silently piling up through the millions upon millions of years. . . .

The saltiest ocean water in the world is that of the Red Sea, where the burning sun and the fierce heat of the atmosphere produce such rapid evaporation that the salt content is 40 parts per thousand. The Sargasso Sea, an area of high air temperatures, receiving no inflow of river water or melting ice because of its remoteness from land, is the saltiest part of the Atlantic, which in turn is the saltiest of the oceans. The polar seas, as one would expect, are the least salty, because they are constantly being diluted by rain, snow, and melting ice. Along the Atlantic coast of the United States, the salinity range from about 33 parts per thousand off Cape Cod to about 36 off Florida is a difference easily perceptible to the senses of human bathers.

Ocean temperatures vary from about 28°F. in polar seas to 96° in the Persian Gulf, which contains the hottest ocean water in the world. To creatures of the sea, which with few exceptions must match in their own bodies the temperature of the surrounding water, this range is tremendous, and change of temperature is probably the most important single condition that controls the distribution of marine animals. . . .

The compression of the sea under its own weight is relatively slight, [but] . . . nevertheless the weight of sea water—the pressing down of miles of water upon all the underlying layers—does have a certain effect upon the water itself. If this downward compression could suddenly be relaxed by some miraculous suspension of natural laws, the sea level would rise about 93 feet all over the world. This would shift the Atlantic coastline of the United States westward a hundred miles or more and alter other familiar geographic outlines all over the world. . . .

As long as there has been an earth, the moving masses of air that we call winds have swept back and forth across its surface. And as long as there has been an ocean, its waters have stirred to the passage of the winds. Most waves are the result of the action of wind on water. There are exceptions, such as the tidal waves sometimes produced by earthquakes under the sea. But the waves most of us know best are wind waves.

It is a confused pattern that the waves make in the open sea—a mixture of countless different wave trains, intermingling, overtaking, passing, or sometimes engulfing one another; each group differing from the others in the place and manner of its origin, in its speed, its direction of movement; some doomed

The land is sculptured by the sea—Grotto of the Pigeons, Beirut, Lebanon. (Gardner Collection, Harvard University.)

never to reach any shore, others destined to roll across half an ocean before they dissolve in thunder on a distant beach. . . .

For millennia beyond computation, the sea's waves have battered the coastlines of the world with erosive effect, here cutting back a cliff, there stripping away tons of sand from a beach, and yet again, in a reversal of their destructiveness, building up a bar or a small island. Unlike the slow geologic changes that bring about the flooding of half a continent, the work of the waves is attuned to the brief span of human life, and so the sculpturing of the continent's edge is something each of us can see for ourselves. . . .

The permanent currents of the ocean are, in a way, the most majestic of her phenomena. Reflecting upon them, our minds are at once taken out from the earth so that we can regard, as from another planet, the spinning of the globe, the winds that deeply trouble its surface or gently encompass it, and the influence of the sun and the moon. For all these cosmic forces are closely linked with the great currents of the ocean, earning for them the adjective I like best of all those applied to them—the planetary currents.

Since the world began, the ocean currents have undoubtedly changed their courses many times (we know, for example, that the Gulf Stream is no more than about 60 million years old); but it would be a bold writer who would try to describe their pattern in the Cambrian Period, for example, or in the Devonian, or in the Jurassic. So far as the brief period of human history is concerned, however, it is unlikely that there has been any important change in the major patterns of oceanic circulation. The first thing that impresses us about the currents is their permanence. This is not surprising, for the forces that produce the currents show little disposition to change materially over the eons of earthly time. The primary driving power is supplied by the winds; the modifying influences are the sun, the revolving of the earth ever toward the east, and the obstructing masses of the continents.

The surface of the sea is unequally heated by the sun; as the water is warmed it expands and becomes lighter, while the cold water becomes heavier and more dense. Probably a slow exchange of polar and equatorial waters is brought about by these differences, the heated water of the tropics moving poleward in the upper layers, and polar water creeping toward the equator along the floor of the sea. But these movements are obscured and largely lost in the far greater sweep of the wind-driven currents. The steadiest winds are the trades, blowing diagonally toward the equator from the northeast and southeast. It is the trades that drive the equatorial currents around the globe. On wind and water alike, as on all that moves, be it a ship, a bullet, or a bird, the spinning earth exerts a deflecting force, turning all moving objects to the right in the Northern Hemisphere and to the left in the Southern. Through the combined action of these and other forces, the resulting current patterns are slowly circulating eddies, turning to the right, or clockwise, in the northern oceans, and to the left or counterclockwise in the southern. There are exceptions, and the Indian Ocean, which seems never to be quite like the others, is an important one....

The conflict between opposing water masses may, in places, be one of the most dramatic of the ocean's phenomena. Superficial hissings and sighings, the striping of the surface waters with lines of froth, a confused turbulence and boiling, and even sounds like distant breakers accompany the displacement of the surface layers by deep water. As visible evidence of the upward movement of the water masses, some of the creatures that inhabit the deeper places of the sea may be carried up bodily into the surface, there to set off orgies of devouring and being devoured, such as Robert Cushman Murphy witnessed one night off the coast of Colombia from the schooner *Askoy*. The night had been still and dark, but the behavior of the surface made it clear that deep water was rising and that some sort of conflict was in progress among opposing water masses far below the ship. All about the schooner small, steep waves leaped into being and dissolved in foaming whitecaps, pricked with the blue fire of luminescent organisms. . . .

When upwelling takes place along coastlines, it is the result of the interplay of several forces—the winds, the surface currents, the rotation of the earth, and the shape of the hidden slopes of the continents' foundations. When the winds, combined with the deflecting effect of rotation, blow the surface waters offshore, deep water must rise to replace it.

Upwelling may occur in the open sea as well, but from entirely different causes. Wherever two strongly moving currents diverge, water must rise from below to fill the place where the streams separate. . . .

The downward movement of surface water into the depths is an occurrence as dramatic as upwelling, and perhaps it fills the human mind with an even greater sense of awe and mystery, because it cannot be seen but can only be imagined. At several known places the downward flow of enormous quantities of water takes place regularly. This water feeds the deep currents of whose courses we have only the dimmest knowledge. We do know that it is all part of the ocean's system of balances, by which she pays back to one part of her waters what she had latterly borrowed for distribution to another. . . .

There is, then, no water that is wholly of the Pacific or

wholly of the Atlantic, or of the Indian or the Antarctic. The surf that we find exhilarating at Virginia Beach or at La Jolla today may have lapped at the base of antarctic icebergs or sparkled in the Mediterranean sun, years ago, before it moved through dark and unseen waterways to the place we find it now. It is by the deep, hidden currents that the oceans are made one.

There is no drop of water in the ocean, not even in the deepest parts of the abyss, that does not know and respond to the mysterious forces that create the tide. No other force that affects the sea is so strong. Compared with the tide, the wind-created waves are surface movements felt, at most, no more than a hundred fathoms below the surface. So, despite their impressive sweep, are the planetary currents, which seldom involve more than the upper several hundred fathoms. The masses of water affected by the tidal movement are enormous, as will be clear from one example. Into one small bay on the east coast of North America—Passamaquoddy—two billion tons of water are carried by the tidal current twice each day; into the whole Bay of Fundy, 100 billion tons.

Here and there we find dramatic illustration of the fact that the tides affect the whole ocean, from its surface to its floor. The meeting of opposing tidal currents in the Strait of Messina creates whirlpools (one of them is Charybdis of classical fame) which so deeply stir the waters of the strait that fish bearing all the marks of abyssal existence, their eyes atrophied or abnormally large, their bodies studded with phosphorescent organs, frequently are cast up on the lighthouse beach, and the whole area yields a rich collection of deep-sea fauna for the Institute of Marine Biology at Messina.

The tides are a response of the mobile waters of the ocean to the pull of the moon and the more distant sun. In theory, there is a gravitational attraction between every drop of sea water and even the outermost star of the universe. In practice, however, the pull of the remote stars is so slight as to be obliterated in the vaster movements by which the ocean yields to the moon and the sun. Anyone who has lived near tidewater knows that the moon, far more than the sun, controls the tides. He has noticed that, just as the moon rises later each day by fifty minutes, on the average, than the day before, so, in most places, the time of high tide is correspondingly later each day. And as the moon waxes and wanes in its monthly cycle, so the height of the tide varies. Twice each month, when the moon is a mere thread of silver in the sky, and again when it is full, we have the strongest tidal movements—the highest flood tides and the lowest ebb tides of the lunar month. These are called the spring tides. At these times, sun, moon, and earth are directly in line and the pull of the two heavenly bodies is added together to bring the water high on the beaches, and send its surf leaping upward against the sea cliffs, and draw a brimming tide into the harbors so that the boats float high beside their wharfs. And twice each month, at the quarters of the moon, when sun, moon, and earth lie at the apexes of a triangle, and the pull of sun and moon are opposed, we have the moderate tidal movements called the neap tides. Then the difference between high and low water is less than at any other time during the month. . . .

The tides present a striking paradox, and the essence of it is this: the force that sets them in motion is cosmic, lying wholly outside the earth and presumably acting impartially on all parts of the globe, but the nature of the tide at any particular place is a local matter, with astonishing differences occurring within a very short geographic distance. . . . The tidal rhythms, as well as the range of tide, vary from ocean to ocean. . . . There seems to be no simple explanation of why. . . . Depending on local geographic features, every part of earth and sea, while affected in some degree by each component of the tide, is more responsive to some than to others. . . .

The range of the tides is measured in tens of feet, and mariners are greatly concerned not only with the stages and the set of the tidal currents, but with the many violent movements and disturbances of the sea that are indirectly related to the tides. Nothing a human mind has invented can tame a tide rip or control the rhythm of the water's ebb and flow, and the most modern instruments cannot carry a vessel over a shoal until the tide has brought a sufficient depth of water over it. . . .

The influence of the tide over the affairs of sea creatures as well as men may be seen all over the world. The billions upon billions of sessile animals, like oysters, mussels, and barnacles, owe their very existence to the sweep of the tides, which brings them the food that they are unable to go in search of. By marvelous adaptations of form and structure, the inhabitants of the world between the tide lines are enabled to live in a zone where the danger of being dried up is matched against the danger of being washed away, where for every enemy that comes by sea there is another that comes by land, and where the most delicate of living tissues must somehow withstand the assault of storm waves that have power to shift tons of rocks or to crack the hardest granite. . . .

For the globe as a whole, the ocean is the great regulator, the great stabilizer of temperatures. It has been described as "a savings bank for solar energy, receiving deposits in seasons of excessive insolation and paying them back in seasons of want." Without the ocean, our world would be visited by unthinkably harsh extremes of temperature. For the water that covers three-fourths of the earth's surface with an enveloping mantle is a substance of remarkable qualities. It is an excellent absorber and radiator of heat. Because of its enormous heat capacity, the ocean can absorb a great deal of heat from the sun without becoming what we would consider "hot," or it can lose much of its heat without becoming "cold."

Through the agency of ocean currents, heat and cold may be distributed over thousands of miles. It is possible to follow the course of a mass of warm water that originates in the trade-wind belt of the Southern Hemisphere and remains recognizable for a year and a half, through a course of more than 7,000 miles. This redistributing function of the ocean tends to make up for the uneven heating of the globe by the sun. As it is, ocean currents carry hot equatorial water toward the poles and return cold water equator-ward by such surface drifts as the Labrador Current and Oyashio, and even more importantly by deep currents. The redistribution of heat for the whole earth is accomplished about half by the ocean currents, and half by the winds.

At that thin interface between the ocean of water and the ocean of overlying air, lying as they do in direct contact over by far the greater part of the earth, there are continuous interactions of tremendous importance.

The atmosphere warms or cools the ocean. It receives vapors through evaporation, leaving the salts in the sea and so increasing the salinity of the water. With the changing weight of that whole mass of air that envelops the earth, the atmosphere brings variable pressure to bear on the surface of the sea, which is depressed under areas of high pressure and springs up in compensation under atmospheric lows. With the moving force of the winds, the air grips the surface of the ocean and raises it into waves, drives the currents onward, lowers sea level on windward shores, and raises it on the lee shores.

But even more does the ocean dominate the air. Its effect on the temperature and humidity of the atmosphere is far greater than the small transfer of heat from air to sea. It takes 3,000 times as much heat to warm a given volume of water 1° as to warm an equal volume of air by the same amount. The heat lost by a cubic meter of water on cooling 1°C. would raise the temperature of 3,000 cubic meters of air by the same amount. Or, to use another example, a layer of water a meter deep, on cooling .1° could warm a layer of air 33 meters thick by 10°. The temperature of the air is intimately related to atmospheric pressure. Where the air is cold, pressure tends to be high; warm air favors low pressures. The transfer of heat between ocean and air therefore alters the belt of high and low pressure; this profoundly affects the direction and strength of the winds and directs the storms on their paths.

Navigators sailed the seas for centuries without any
idea of the depths of water under them except in harbors.
When Magellan failed to touch bottom in the Pacific with
a weight on 1,200 feet of rope, he announced erroneously
that he was over the deepest part. The first true deep-sea
sounding was in 1840 by Sir James Clark Ross, who
let out 12,000 feet of hemp line before reaching bottom.
In the following years, details on ocean depths slowly
accumulated as measurements were made by lowering
weights to the bottom. However, as late as 1923 our
information was still scanty and unreliable. Then echo
sounding came into use, and ships were able to make
continuous records of the depth of water under them by
sending sound waves into the water and measuring the
time required for them to echo from the bottom and
return to the ship. Since then, we have learned much.

An unexpected discovery was that the bottom of the
deep sea is not the flat monotonous plain it was once
thought to be. Its surface is ridged with mountain ranges
and deep trenches, and pocked with volcanic peaks.

Interest grew in finding the deepest of the trenches,
and echo soundings indicated that it was in the trough
that borders the Mariana Islands of the western Pacific.
But echo soundings are indirect ways of locating the
bottom, and there is always the challenge to **see** where
the deeps are.

On January 23, 1960, two men went down for a look at Challenger Deep, in the United States Navy bathyscaph **Trieste.** They were Lt. Don Walsh of the United States Navy, and Jacques Piccard, son of the Swiss scientist Auguste Piccard who designed the **Trieste.** They went down nearly 7 miles, by far the greatest depth ever penetrated by man. Down there, even within the limited range of their searchlight, they saw a flounderlike fish browsing, and a shrimp, bright red and about an inch long, floated by.

The following article is from the story of the dive told by Lieutenant Walsh in **Life Magazine.**

Don Walsh

THE OCEAN'S DEEP

Three things always happen on the Trieste *when she's on her* way, and all of them happened at once today: the needle on our sensitive depth gauge began to quiver downward, the rocking motion of the sphere became perceptibly less violent and—why this happens we have never been able to figure out—the stern settled by a degree or so. . . .

After we had been under way for four minutes I called the *Lewis* on the underwater telephone and reported that we were all right and passing 250 feet. I also told them about losing our topside telephone and the current meter. By the time I had finished, we were at 300 feet.

At that point we encountered the thermocline, a layer where the water temperature drops sharply. Since the cold water was denser than the water we had been passing through, we became relatively more buoyant and stopped. We had expected this. Part of our standard diving procedure is to use this brief halt as an opportunity to make a final instrument check. Then, by releasing a little gasoline from our maneuvering tank, we get rid of some of our excess buoyancy and start down again.

But this thermocline appeared to be a little different from most. We released the gasoline, but the temperature differential was so great that for a moment our depth gauge reported that we

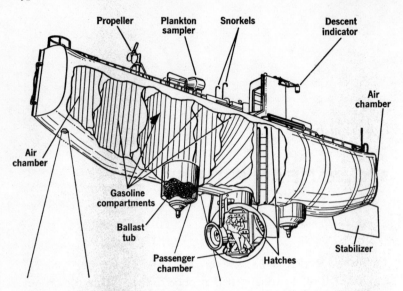

The bathyscaph operates basically like a balloon. It consists of a buoyant tank with a passenger chamber attached to the bottom. A balloon is buoyant when filled with lighter-than-air gas; the bathyscaph tank is buoyant when filled with lighter-than-water gasoline. When divers want to descend, they release gasoline, just as aeronauts release gas. To rise, they drop ballast. The passenger chamber is heavily built to withstand tremendous pressures. The tank is lightly built, with an opening in the bottom through which sea water enters, equalizing pressure inside and outside as the craft descends. Propellers on top are battery-powered and enable the bathyscaph to maneuver at the bottom of the sea. (After Life Magazine February 15, 1960.)

were actually moving back toward the surface. It took another release of gasoline to get us moving down once more. But we were not yet through with the thermocline: we hit it at 400 feet, again at 490 feet and finally at 550 feet. Jacques said he had never in any of his dives encountered such a difficult temperature barrier. We thought perhaps the heavy winds of the preceding days had abnormally shuffled the water layers.

It had taken us more than 20 minutes to fight through the 250-foot thermocline. By contrast we needed only 12 minutes to go from the 1,000-foot to the 2,000-foot level.

At about 600 feet we entered a zone of deepening twilight where colors faded off into gray. By 1,000 feet the light had gone completely. We turned out the lights in the sphere to watch for the luminescent creatures that are sometimes visible at this level. We saw very few. Eventually we turned the cabin lights back on and briefly tested the forward lights that throw a beam in front of the observation window. Formless plankton streamed past, giving us a sensation of great speed.

We were now dropping fast, at about four feet per second. It was getting colder in the sphere, and we decided to put on dry clothing. It was quite an operation to see: two grown men changing clothes in a space 38 inches square and only five feet, eight inches high.

Then we ate our first chocolate bars. There was sort of a joke about this between Jacques and me. On the last dive he had brought the chocolate, so I told him that I would buy the lunch for this one, and I had 15 bars put aboard before we left Guam.

During this time there had been little conversation. Friends often ask what we do along the way, how we keep from becoming bored, what we talk about. The fact is that most of the time we are both too busy either to be bored or to talk very much. There are too many instruments to watch, too many adjustments to make.

Then there are minor incidents such as the small leak that always develops in one of the hull connectors—a place where wires from lights and instruments on the outside of the sphere pass through the hull to the recording apparatus inside. The leak starts at about 10,000 feet. It is an old friend, a tiny drip, drip, drip. I timed the drips and found no change from before, which meant that it had not become more serious. We expected it to disappear at about 15,000 feet, when the water pressure packed the plastic sealer in more tightly—and it did.

Up to this point we had managed to maintain voice contact with the people on the *Wandank*, using Project Nekton's specially developed underwater telephone. But now, at 15,000 feet, we lost them—possible because they were a good distance from us laterally as well as vertically. We were truly on our own now except for a crude system of tone signals Larry and I had arranged. By means of a special key the underwater telephone can send out a tone that sounds something like a radio time signal. These carry farther than voice transmission. In our code, all even numbered signals are for good news: two means all is well, four means we are on the bottom, six means we are on the way up. The bad messages come in odd numbers: three means we are having mechanical difficulty and are coming up but not in distress, five means something has gone wrong and we are coming up in an emergency. So far, we have never had to use the odd numbers.

Now and then as we descended I would dictate an observation or instrument reading into the tape recorder. At 18,600 feet and again at 24,000 I called Jacques' attention to the depth gauge, noting that we were surpassing the previous record dives made during Project Nekton. He grinned and waved his hand.

The inside of the sphere is never really silent on a dive. I could hear the characteristic deep hum of the transformers in the electrical equipment. The air regeneration apparatus hissed continuously as fresh oxygen was automatically fed into the air. Beside me on a hook the headphones of the underwater telephone, volume turned up, were crackling with a sound like radio static.

At 27,000 feet we checked our rate of descent to two feet per second by dumping some shot ballast. We were not too sure of the underwater currents here and we did not want to go crashing into a wall of the trench by mistake. As we neared 30,000 feet I started thinking about the changes we had planned to make when we got within 1,000 feet or so of the bottom—which we now were expecting to find only another 3,500 feet below us. I was running through a mental checklist when we heard and felt a powerful, muffled crack. The sphere rocked as though we were on land and going through a mild earthquake.

We waited anxiously for what might happen next. Nothing did. We flipped off the instruments and the underwater telephone so that we could hear better. Still nothing happened. We switched the instruments back on and studied the dials that would tell us if something critical had occurred. No, we had our equilibrium and were descending exactly as before.

We once had a similar experience. During a descent to 24,000 feet a running lamp that someone had forgotten to remove and a length of deck tubing had imploded—had blown in under the tremendous pressure. The shock waves then had given us a scare, but the performance of the bathyscaph itself hadn't been affected. Something of the same sort seemed to be the case here. Without a formal discussion on the point, we agreed to go on.

We dumped more ballast, checking our speed to one foot per second. At 33,000 feet, only about 600 feet off the expected bottom, we turned on our sensitive Fathometer, which always before had quickly and accurately picked up the floor for us. It showed nothing. We continued to slide on down. It also showed nothing 100 feet later—or 100 feet below that. At 36,000 feet Jacques asked me wryly whether I thought we could have missed the floor somehow. I said I thought the chances were against it.

We checked our speed to half a foot a second and continued. At that rate, time and distance pass very slowly, and I think for the first time in the dive both of us had the feeling of awe that comes from exploring the totally unknown.

I did not take my eyes off the Fathometer and Jacques never stopped watching out of the tiny porthole with its weak probe of light. No bottom was in sight at 36,600 feet, or at 37,200. But at last at 37,500 feet the Fathometer traced the beginnings of the bottom. Soon Jacques could see a difference in the effect of our light in the water, as the rays reflected off the bottom. As we approached the floor I called the Fathometer readings to Jacques in fathoms: "Thirty . . . twenty . . . ten . . ." At eight, he called that he could see the gray-white bottom.

As we sank through the clear water near the bottom, we had a tremendous piece of luck. Peering through the tiny porthole, Jacques spotted a fish. It appeared to be browsing, searching for food along the ocean floor. It looked like a sole or flounder, flat with eyes on the side of its head. It was about a foot long. Our sudden appearance in his domain, with our great light casting illumination such as he had never seen before, did not seem to bother him at all. After we had been watching him for a minute, he swam slowly off into the darkness again, beyond the range of our light. It was an exciting event. The fish was obviously a bottom-feeder, which means that it must spend all its life at these tremendous depths, under these tremendous pressures.[1]

At 1:10 P.M. we sank gently onto the soft floor. A great cloud of silt rose around us. We had found the bottom at 37,800 feet, 1,600 feet deeper than the deepest soundings ever made and 4,200 feet deeper than the rough soundings made on the *Lewis* had led us to expect. Obviously, the present method of making soundings is inaccurate and we have a good deal to learn about how fast sound travels through water at those depths. (Our depth figure has not yet been corrected. When it is, it probably will be found to be a trifle less than 37,800 feet.)[2]

The silt cloud was still around us, so for the moment we could make no more visual observations. I keyed the underwater telephone four times, the signal that we had reached the bottom. Then, with no expectation that I would be heard, I called on the voice circuit: "*Wandank, Wandank,* this is *Trieste.* We are on the bottom of the Challenger Deep at 63 hundred fathoms. Over." To our complete astonishment Larry's voice came back: "*Trieste, Trieste,* this is *Wandank.* I hear you faint but clear. Will you repeat your present depth? Over." I didn't wonder that he wanted to make sure he had heard our depth correctly. I gave it to him again and back came Larry. We could sense the excite-

---

[1] At these depths pressure is 18,000 pounds per square inch. Surface pressure is 14.5 pounds per square inch—editors' note.

[2] When the instruments were calibrated later, the depth was corrected to 35,800 feet (6.8 miles)—editors' note.

ment in his voice. "*Trieste*, this is *Wandank*. Understand six three zero zero fathoms. Roger. Out."

Solemnly Jacques and I shook hands. Then he unrolled a Swiss flag he had brought along, and I unrolled a U.S. flag and we took pictures of ourselves with an automatic camera belonging to *Life*.

While we waited for the sediment to settle down, Jacques had a quick view of the second and last piece of animal life we were to encounter. What seemed to be a small shrimp, bright red and perhaps an inch long, floated by in the middle of the mud cloud. We were elated. To have seen not just one but two live creatures at the bottom, especially with the equipment we had, was staggeringly lucky. It was the equivalent of seeing a rare animal while sealed in a small steel ball on top of Mt. Everest for 20 minutes in the middle of the night with no means of illumination but a flashlight fixed to the side of the sphere.

After about 10 minutes the bottom began to be visible again, and I decided to look at it through the after porthole. The primary purpose of this porthole is to give us a view of one of the tubs containing ballast so that we can check the amount visually. You look through the porthole and then through a Plexiglass window set in the bottom of the antechamber. I switched on the light and looked out.

I saw the ocean floor, which looked flat, but I also saw what had jarred us at 30,000 feet. Across the Plexiglass window ran a series of cracks, stretching from one side to the other.

This was worrisome. The crack presented no threat to our safety, but it did make it questionable how easily we would be able to get out of the bathyscaph once we reached the surface. Our procedure when we get up is to blow the water out of the long passageway up to the conning tower, using an arrangement of tubes and valves and bottles of compressed air. But if this cracked window should shatter, it would let the water back in— and we would be stuck in the sphere. We had a snorkel arrangement that gave access to the outside, so once we were back at the surface we would at least have air and probably be able to get liquids too. But we would have to stay shut in for four or five days while they towed us back to Guam, took the gasoline out of the float, hauled the whole bathyscaph out of the water and drained the passageway.

This would be highly uncomfortable, and we decided we had better get to the surface as quickly as we could—we had been on the bottom for 20 of the planned 30 minutes anyhow. We wanted to give the people on the *Wandank* and the *Lewis* as much daylight as possible to find us, rig the tow and perhaps even send a diver down to see if some quick repair could be made. So we dumped two tons of ballast, signaled we were coming, and started up. The trip to the surface took three hours and 27 minutes—an hour and 11 minutes less than the trip down. But it seemed much longer. We had a feeling of anticlimax. The big moment had passed. We knew *Trieste's* achievement had been extraordinary, but all we could think of was whether we would have to spend four more days cooped up in this sphere.

We were both very cold now. The temperature at the bottom had been 37.4°F. and the temperature in the sphere was a damp 45°. There is no air circulation in the sphere, so the cold air, being heavier, accumulates near the floor. Our feet were especially cold.

We ate several more chocolate bars on the way up, but not as many as we really wanted. Jacques argued that we ought to save as many as possible in case it turned out that we were going to be in there for a few days.

As we rose, we saw a curious thing. Mud, which had adhered to the bottom of the sphere when we pulled up, was apparently caught in a back eddy created by the uprushing sphere. It now flowed upward past our window, giving us the illusion that we were going down again. Mixed with the mud were flecks of paint from the sphere itself. At the depth we reached, the pressure is almost nine tons per square inch. This is enough to compress the sphere by two millimeters, which loosens small flecks of paint.

We reached the surface at 4:57 P.M. Now we had to find out just exactly where we were going to spend the next few days.

Usually at the end of a dive we are impatient. We blow the passageway quickly and violently. This time we tried it slowly and gently so as not to jar the cracked window. I fed first one, then a second, then a third bottle of compressed air into the system as Jacques watched for results through the hatch window. Finally, ever so slowly, the water level moved down past the window. A stream of bubbles broke from the air line, meaning that the water was gone from the chamber.

We wasted very little time opening the big hatch and climbing out. We closed it carefully behind us again and hurried up the long ladder, opened the conning-tower hatch and came out into the sunshine and that wonderful fresh air.

We seemed to have arrived on the surface in the middle of a carnival. Two navy jet photo planes blasted by us a few feet above the conning tower. An Air Force search and rescue C-54 made several low passes at us. Off to the west, the *Lewis* was bearing down on us and right behind her came the *Wandank*, breaking all speed records for tugs. She looked like a 30-knot destroyer heading for a fire. She looked great. So did everything else.

What did the record-breaking dive of the *Trieste* accomplish?

The descent into the Challenger Deep was the carefully planned, logical climax of years of work, first by the Piccards and later by our Navy. The achievement of the final great depth was intended to show that man now can personally investigate any part of any ocean. The conquest of the Challenger Deep was the ultimate test of the bathyscaph as a research platform.

Research into the deepest oceans is becoming increasingly important. The feat of the *Trieste* puts the U.S. on the threshold of a new era in Oceanography and demonstrates the role of the Navy in its advancement. If the oceans' depths are to be the potential battlegrounds of the future, we must learn all we can about them. The bathyscaph's descent was the greatest advance to date in this effort.

A science is merely the organized compilation of observations and theories made by people. An important part of understanding it is knowing something about the people who contributed to it, how and why they did what they did. Geology comprises the work of so many people that it is difficult, if not impossible, to identify any one individual with major advances in a particular area. But from the past we have examples of a few individuals who nearly single-handed guided thinking into channels that led to spectacular progress. Louis Agassiz was such a man.

The story of Louis Agassiz and the growth of the glacial theory shows how one man became a geologist almost accidentally and made one of the outstanding contributions in the history of the science. From the first, he was curious about nature. But the qualities that threaded through his life and made him outstanding in his ultimate work, in addition to habitual curiosity, were patient thoroughness and perseverance in all he did, whether it was helping a cooper tighten casks, making leather shoes for his sister's dolls, hand-copying books he could not afford to buy, or studying medicine for a career he never followed.

As often happens in science, the ideas about glaciers for which Agassiz became famous did not actually originate with him. They started from a suggestion made by an observant and thoughtful mountaineer that certain gravel ridges and trains of huge boulders in Alpine valleys were left by glaciers which once spread far beyond where they were standing early in the nineteenth century. Agassiz's first investigations into the subject were for the purpose of demonstrating how foolish the idea was. Though he went to scoff, he quickly admitted the compelling evidence his eyes found. In the years following his own acceptance of the evidence, he marshaled and described observations in the Alps, Great Britain, and Scandinavia that finally convinced many of his doubting contemporaries. When he traveled to Boston in 1846 to lecture at the Lowell Institute, he found many of the unique signs of past glaciation in eastern Massachusetts. After he had decided to remain in the United States to teach at Harvard, he extended the scope of his investigations to the Lake Superior region, the Finger Lakes of New York, and even the Straits of Magellan when he was en route to San Francisco by way of Cape Horn. He blundered, too, as any of us might. He misinterpreted some rock outcrops in Brazil that had weathered to an unusual shape as having been shaped by glaciers. There he generalized too broadly from narrowly limited evidence.

The story of Louis Agassiz is the story of one geologist, but it could be the story of anyone with a driving curiosity and a willingness to work hard at anything he does. Beyond these qualities, he had no unique or mysterious talents.

The following sketch of Agassiz's life is by Carroll Lane Fenton and Mildred Adams Fenton, among the best-known writers of popular works in geology of our day, from their book **Giants of Geology.**

Carroll Lane Fenton and Mildred Adams Fenton

A GIANT OF GEOLOGY: AGASSIZ

Jean Louis Rodolphe Agassiz was born on May 28, 1807, at Motier on the shores of the Lake of Morat, in west central Switzerland. There his parents had moved after years in the bleak village of St. Imier, where the elder Agassiz preached confidence in God to peasants who needed proofs of divine beneficence after labor in their rocky fields. In such surroundings pastor was little better off than peasant and both approached poverty when crops were poor or dairy herds met disaster. To these hardships was added death, which took child after child till both the pastor and his wife gave up. After praying over a fourth small grave the couple packed their possessions and moved.

Motier was neither large nor rich, but it seemed both to the Rev. Agassiz and his wife. Soil of the farms was deep and good; the parsonage was substantial; its vineyard promised an income in addition to salary. Garden and orchard provided food, while a spring brought plentiful water to a stone pool just behind the house. That pool formed the aquarium in which small Louis kept fish when he learned to catch them in near-by shallows of the lake.

For the boy became a naturalist almost in infancy. He began

with rabbits and guinea pigs, but in time reared field mice, snakes and even wild birds in cages under trees of the yard. In fishing he used his hands, treating captives so carefully that they swam and ate vigorously when he transferred them to the stone basin. There the boy sat with his nose close to the water, watching their movements as he noted their shapes, their colors, and the delicate bones in their fins.

Another quality appeared early; that of leadership. Louis guided his playmates and urged them to adventure; his younger brother followed him in both natural history and games. When Louis went fishing so did Auguste; when Louis wanted to skate across the lake Auguste came without hesitation. A crack in the ice seemed too wide to cross, but when Louis turned himself into a bridge his brother crept over it without fear. The mother spied them through a telescope and in terror sent a workman on skates to bring the youngsters home. He led them back across the lake, and Louis was the one who protested. If the ice was unsafe for two small boys, how could it be less dangerous for two boys and a man? And besides, why not go to town and ride home in comfort with Father?

Until Louis was ten he studied at home, learning to read, write and figure with Pastor Agassiz as his teacher. He also studied less formally under workmen—the cobbler who came twice a year, the tailor, the carpenter, and the cooper who tightened old casks or made new ones before grapes were ready to harvest. When eight or nine years old the boy could make pens for his pets and could cut and sew leather shoes for Sister Cecile's dolls. He also knew how to mend his own torn clothing and make a model barrel that was watertight. As a man he would credit his skill in dissection to this early training in crafts.

At ten Louis entered the college, or school, for boys at Bienne; Auguste, true to form, followed a year later. Discipline was strict and work was hard: nine hours of study per day, with mathematics, Greek, Latin and modern languages. But there were frequent recesses for games, and home was only twenty miles away. When vacation came the boys made this trip on foot, saving money to buy inexpensive books which Louis always selected. Sometimes they went to Cudrefin instead of Motier for a stay with their mother's father, the old and respected Dr. Mayor. He let them drive his small white horse or come to Easter celebrations where vast numbers of eggs and fritters were eaten and a special dance was announced in honor of the Mayors. Only the family friends and some neighbors took part while other villagers looked on.

Four years Louis Agassiz spent at Bienne, and then drew up a plan for his future education. It expressed willingness to serve eighteen months of apprenticeship in commerce, after which he was determined to "advance in the sciences" and "become a man of letters." Such a program meant immediate studies in Greek, Latin, Italian, and geography both ancient and modern, with books that would cost twelve louis. Apprenticeship would be followed by four years at a German university, after which final studies in Paris would consume about five years. Then, at the age of twenty-five, he could begin to write.

An ambitious plan for the son of a village pastor who received much of his salary in food. Louis broached it first to Grandfather Mayor, who was impressed by the boy's notebooks and letters of recommendation from the faculty at Bienne. But there was, he knew, the problem of money; of money to be spent then with care, and of more that must be earned as Louis became a man. Why not take up medicine, which could furnish a reliable income?

Because, began Louis—and then stopped. Medicine, after all, was a science close to zoology and vastly more attractive than business. Grandfather would take his part, and so would Uncle Matthias, a noted physician of Lausanne. There was an evening filled with examination of more notebooks, re-reading of letters, and careful budgeting. In the end it was decided that Louis should continue his college work at Lausanne, with instruction in anatomy under his famous uncle. Since Auguste was too young for business he would go along.

For two years the boys stayed at Lausanne, turning their

room into a zoo and delighting a professor who had charge of the canton museum. Then Louis entered the medical school at Zurich, still with Auguste in his wake. The boys lived in a private house, made one inspiring trip to the Alps, and spent many hours copying books which they could not afford to buy. Two years of this and they finally parted, Auguste to enter business while his brother went on to Heidelberg and a medical degree.

Louis Agassiz was not quite nineteen when he entered Germany in April 1826. A pastel sketch shows him as a wavy-haired youth with thin mustache and a wide, open collar. He rose at six o'clock, went to lectures at seven, and kept busy until nine at night. He was disturbed by a charge of six crowns for matriculation, pleased with his new professors, and happy to meet a student of botany named Alexander Braun. They soon became inseparable, and when Agassiz fell ill of typhoid he was taken to the Braun home in Carlsruhe. He left it with a pledge to marry Alexander's sister, Cecile.

Illness was followed by several months at home, during which Agassiz studied the growth of tadpoles, collected fish from mountain lakes, and gathered specimens of plants. In October 1827 he joined Braun in a "pilgrimage" to Munich, where lectures were free, board was cheap, and beer good as well as plentiful. On the way they saw their first llama, stuffed and in a museum near the skeleton of a mammoth, which then was called a *carnivorous* elephant.

Pastor Agassiz had described his son as "courageous, industrious, and discreet"; a youth who "pursues honorably and vigorously his aim, namely, the degree of Doctor of Medicine and Surgery." Yet honor did not preclude detours, and by the summer of 1828 Louis was busy writing a book about fishes which his Professor von Martius had brought from Brazil. The work was in Latin, with forty colored plates, and to grace its title page the author became a doctor of philosophy. He received the M.D. degree on April 3, after nine days of examination. To his parents it was a "most precious laurel" which assured their son of a career "as safe as it was honorable."

At twenty-two Louis Agassiz cared nothing for safety, nor did he want a medical career. He had begun a great book on fossil fishes and another describing those of lakes and streams. With an allowance of two hundred and fifty dollars per year he supported himself, traveled to study specimens, and paid an artist to draw them. On borrowed money he reached Paris in 1831, where the foremost scientists entertained him in spite of his poverty. A year later he became professor of natural history in the new college at Neuchâtel, just across the lake from Grandfather Mayor's home. The institution had fewer than a hundred students, little money and no buildings; Agassiz received only eighty louis (about four hundred dollars) per year and lectured in the city hall. He had to start his museum in a refuge for orphans and turn the home to which he brought his bride into a combination of laboratory and boardinghouse in which two assistants, a collector and an errand boy, helped him with his work. The collector brought in fossils which no one else could find, but he also slept in his clothes and seldom changed them. He undoubtedly helped convince the young Mrs. Agassiz that people of Neuchâtel were less pleasant than those of her native Carlsruhe.

The collector's fossils were stuff for another big book, but before it could be written Agassiz found a new enthusiasm. In the summer of 1836 he took his wife and baby son to Bex, where Cecile could relax with Mme de Charpentier, German wife of an amateur naturalist who managed the local salt works. Charpentier agreed to guide Agassiz on trips through the mountains in search of petrifactions and fish. There also would be visits to gravel ridges and trains of huge boulders supposed to have been left by glaciers that once spread into the central Swiss plain and across the slopes of the Jura.

This supposition was not new with Charpentier; he had received it back in 1815 from a thoughtful mountaineer. For fourteen years the suggestion lay fallow; then it was taken up by a friend who wrote a paper on changing climates of the Alps. Charpentier at last was convinced, as was a learned political refugee from Breslau. But to Agassiz the theory of ancient, long-vanished glaciers seemed nonsense, though he was willing to examine the evidence. Charpentier and he could have grand trips together, even if they found no ancient ice.

Instead of playing the part of skeptic, Agassiz was carried away by the truth of his host's "nonsense." Together they examined the existing glaciers of Diablerets, and Chamonix, where Lyell had worked, followed ridges of drift along the Rhone, and traced those of tributary valleys. Water could not have piled up

A glacial erratic—Cohasset, Massachusetts. This huge granitic boulder weighs many tons and is wholly unlike the underlying rock. It was carried to this point by a glacier, then dropped when the ice melted. (John Shimer.)

such deposits; only moving, melting ice could have left them after scouring and scraping bedrock, at the same time carrying great boulders scores of miles from their source. And if glaciers had done such work near the Rhone, must they not have operated wherever drift, scoured bedrock and "erratic" boulders were found? Agassiz insisted that Charpentier publish his facts, in the hope of stimulating other naturalists to make similar observations. He himself went glacier-hunting in the Jura and, when the Helvetic Association met in 1837, was ready with a presidential address which one critic called a "fiery discourse about a sheet of ice."

One rare quality of Agassiz was his power to see great subjects as wholes; to state large problems and conclusions so comprehensively that his first work commanded respect. He had done this in the generalizations of his preface to *Fossil Fishes*; he now displayed his ability as well as daring by going directly from the evidence for ancient, long-melted glaciers to the fact of a glacial period. That period, he suggested, had begun with a temporary but world-wide and rapid change in climate which allowed a sheet of ice to spread from the North Pole to Central Europe and Asia.

> Siberian winter [he concluded] established itself for a time over a world previously covered with a rich vegetation and peopled with large mammalia, similar to those now inhabiting the warm regions of India and Africa. Death enveloped all nature in a shroud, and the cold, having reached its highest degree, gave to this mass of ice, at the maximum of tension, the greatest possible hardness.

Von Buch was in the audience; his disapproval, tinged with tolerant contempt, bubbled over when the young president stopped speaking. Humboldt, a staunch friend and benefactor, expressed his objections by letter. "Over your and Charpentier's moraines Leopold von Buch rages, as you may already know. . . . I, too, though by no means so bitterly opposed to new views, and ready to believe that the boulders have not all been moved by the same means, am yet inclined to think the moraines due to more local causes." And again, "Your ice frightens me. . . . I am afraid you spread your intellect over too many subjects at once."

Agassiz did not share that fear, nor was he alarmed by ice. Besides his work as professor he was printing and illustrating *Fossil Fishes* and *Fresh-Water Fishes*, as well as books on mollusks and on the group to which sea urchins belong. He maintained a lithographic plant with as many as twenty workmen, publishing books by other authors to keep his employees busy. Yet when August of 1838 came round he was ready for a trip to the glaciers of Mont Blanc, and a year later studied ice streams of Monte Rosa, the Matterhorn, and several other mountains. At one place he examined a cabin built on ice in 1827; in twelve years it had traveled four thousand feet and seemed to be gathering speed.

In 1840 Agassiz published his *Studies on Glaciers*, with one large volume of text and an atlas of thirty-two plates. It reviewed earlier work on the subject and presented new facts about the appearance and structure of these streams of slowly moving ice, their formation and internal temperatures, and the loads of broken and pulverized rock scattered through them or carried on their surface. Most stimulating, however, were chapters dealing with ancient Swiss glaciers and with the ice sheet which had brought arctic conditions to a once-warm continent:

> The surface of Europe, adorned before by a tropical vegetation and inhabited by troops of large elephants, enormous hippopotami, and gigantic carnivora, was suddenly buried under a vast mantle of ice, covering alike plains, lakes, seas, and plateaus. Upon the life and movement of a powerful creation fell the silence of death. Springs paused, rivers ceased to flow, the rays of the sun, rising upon this frozen shore (if, indeed, it was reached by them), were met only by the breath of the winter from the north and the thunders of the crevasses as they opened across the surface of this icy sea.

The main case was presented; now for details. In the summer of 1840 Agassiz hired a mason to wall in a huge block of micaceous schist that stood on a ridge of rock rubbish, or moraine, bisecting the lower Aar glacier. The uneven floor was smoothed with slabs, a blanket served as door, and the whole formed a hut that would shelter six people if none minded his neighbor's elbows. A niche under another boulder served as storehouse for food.

To this hut, the "Hôtel des Neuchâtelois," Agassiz took friends and guides for detailed investigation of the glacier. They carried heavy loads of instruments, as well as an auger for boring holes through which recording thermometers could be lowered into the ice. Microscopes were ready for examination of insects, small plants and other things living upon the moraine. An engineer determined the position of eighteen large boulders which were to be sighted and plotted year after year to determine how fast the ice was moving.

Close-range studies were varied by longer trips on which the whole party climbed such "unscalable" peaks as the Jungfrau and Schreckhorn. Sometimes they went lightly laden; more often their rucksacks were filled with barometers, thermometers, and instruments for simple surveying. On one trip they carried heavy loads through loose snow that came to their knees and then climbed a flight of steps which their guides chopped in the ice. At the summit they began a peasant dance, to stop abruptly as a band of chamois appeared from behind a rock. The trip down was largely a matter of sliding, though they traversed one great crevasse on an ice bridge one to two feet in width and broken near the end. There, one of the climbers noted, "we were obliged to spring across."

Results of this work appeared in the *Glacial System*, published in 1846. Meanwhile Agassiz had been to Britain, where Buckland (Lyell's old teacher) had forsaken deluges for ice. In 1840 Lyell himself was convinced, finding a chain of Ice Age moraines two miles from his father's house, Kinnordy. He and Buckland supported Agassiz at a meeting of the Geological Society of London in November 1840, and though Murchison tried to oppose them, his arguments had little weight. Two years later, Charles Darwin traced ancient glaciers of North Wales and was happy to find a moraine which Buckland had overlooked. The

King of Prussia granted almost a thousand dollars for further work on the Aar, and Agassiz began to plan a trip to North America. But he delayed it to complete the *Researches on Fossil Fishes* as well as a volume on queer fishlike creatures from Devonian sandstones of the British Isles. He also hastened to complete other works, to settle affairs of the college at Neuchâtel, and to close his profitless printing and lithographing plant. No wonder that Humboldt wrote him: "For pity's sake husband your strength!"

At two o'clock on a night in March 1846 Agassiz left Neuchâtel for Paris and the United States. The King had given him fifteen thousand francs for travel; John A. Lowell had engaged him to lecture at the Lowell Institute in Boston. On shipboard Agassiz practiced English; on landing he hastened to survey the country between Boston and Washington. To his mother he wrote enthusiastically about the "frightful" speed of American trains, the hospitality of cultured people, and the boulders, moraines and ice-scoured bedrock of eastern Massachusetts. He also reported the "conversion" of several scientists whose ideas about drift had ranged from hazy to grotesque.

Agassiz gave his double course of English lectures, which were free, and added a subscription series in French which prosperous young ladies of Boston attended with enthusiasm. He left the young ladies to study turtles near Charleston, where jellyfish also were most attractive. By April he was back in New England, leasing a house in East Boston at a high rental partly because it was large and partly because it had a back yard that dipped into the harbor. In this house Agassiz welcomed assistants who had been with him on the Aar and at Neuchâtel, as well as the pastor —now in exile—who had paid for his first trip to Paris. Papa Christinat managed the household while Agassiz went to Niagara Falls, collected fish along the St. Lawrence, and lectured in every large city from Albany back to Charleston.

Agassiz had planned to go back to Switzerland after three years of travel in the United States. But the spring of 1848 found Europe in turmoil; French revolutionaries had proclaimed a re-

Louis Agassiz as he appeared while teaching at Harvard University. (Museum of Comparative Zoology, Harvard University.)

public, Neuchâtel had rebelled against Prussian rule, and several friends besides Papa Christinat were exiled from their homes. Agassiz had no wish to return, but when his Prussian travel grant stopped he had to hunt a job. He was happy to become professor of natural history in the new scientific school at Harvard, with an assured salary of fifteen hundred dollars per year. This sum, guaranteed by the founder, must do until students' fees reached three thousand dollars.

The first course which Agassiz gave at Harvard began in April 1848. At about that time he moved from Boston to Oxford Street, in Cambridge, where there was space for a garden and for living animals. Papa Christinat preferred the former, sighing as space was given to useless turtles and a tank for alligators. A tame bear was chained in one part of the yard; eagles had a cage built close to the wall; smaller cages held a family of opossums and a hutch of rabbits to be used in experiments. There also were vast collections of dead things in a shanty perched on piles above the Charles River.

The collections contained few fossils or rocks, for Agassiz was enthralled by the wealth of undescribed animals to be found in North America and surrounding seas. He studied mammals, turtles, fish, and odd invertebrates that peopled rocky New England coasts. His kindliness and enthusiasm fired pupils, who turned the new house into a laboratory almost as crowded as the one back in Neuchâtel.

Still, the Ice Age could not be forgotten on a continent where ancient glaciers had covered five million square miles. In the summer of 1848 Agassiz took students to Lake Superior, where they traveled in bark canoes through what then was wilderness. As he had done in Switzerland and Great Britain, the professor showed that currents of water could not have scattered boulders without sweeping far beyond the limits in which they were found. Icebergs were inadequate; moreover, they called for a climate cold enough to produce great icecaps. Scratched boulders and bedrock recalled the Alps; ridges and sheets of drift matched those left as Swiss glaciers melted.

Cecile Agassiz had never been reconciled to her husband's impetuous pursuit of science and his unpredictable ways. When he left Neuchâtel she took refuge with her parents in Carlsruhe, where the Agassiz youngsters were happy. Their mother brooded and soon became ill, an illness that ended in death while he was on the trip to Lake Superior. Returning to Cambridge, he assuaged his grief by planning how the children should come to America.

Papa Christinat listened and then attacked. Come to America, indeed! Let Louis look and use his head. Was this a place to bring children—this place which was part boarding house, part laboratory, with men squabbling in every room and a zoo in the back yard? Get rid of those quarrelsome assistants; marry some rich American woman—he knew some who already were preening to catch the professor's eye. Achieve security and a civilized home before sending for the children!

Agassiz could have used wealth, for no scientist ever had more need for money or made fewer efforts to get it. But to marry wealth—that was something else; something not to be done by a man whose ways were as unconventional as those of Agassiz. He shook his head at advice, sent for his sons, and paid court to Elizabeth Carey. She was a professor's sister-in-law and poor, but what matter? Did she not know how to be companion as well as helpmate of a genius? Was she not eager to mother the girls, and did she not capture young Alexander's heart the day he arrived in Cambridge?

Papa Christinat grumbled and moved; Elizabeth and Louis sent for the girls; the bride turned the dirty, orderless Oxford Street house into a cheerful home. By August of 1850 the Agassiz family was united, although burdened with debts that dated back to the frenzied years at Neuchâtel. Elizabeth pinched pennies for five ineffectual years and then, with Alexander's aid, set up a school for girls on the upper floors of the house. She charged high fees but gave thorough instruction, with the advantage of occasional lessons by Harvard's most famous professor. In eight more years all debts were paid and the school could be discontinued.

By that time Agassiz was receiving a worth-while salary, a legacy and state funds had given him a museum, and rich men had made grants for research. In 1865 one wealthy admirer sent the professor, his wife, and several assistants on an expedition to Brazil. They collected fourteen hundred new species of fish, while Agassiz mistakenly reported that glaciers had once advanced to the region of Rio de Janeiro. Here he was misled by weathering that had rounded great blocks of stone where they stood and had smoothed bare hills till they looked like the "sheep rocks" of New England or Switzerland.

Better work was done in 1868, when he crossed wide moraines of the prairies and examined the ice-gouged Finger Lakes of New York. In 1872 he and Mrs. Agassiz sailed around Cape Horn and to San Francisco on a vessel of the Coast Survey. Near the Straits of Magellan they found thick moraines, traced the movements of vanished ice, and examined existing glaciers. Though not so thick as ice streams of Switzerland, some of these were much wider than any on the continent of Europe.

This trip almost completed Agassiz's work; after a school year and one more summer of teaching, he died in 1873. Some conservatives still denied his great Ice Age; other men were tracing sheets of drift as well as intervening deposits which would show that the Glacial Period consisted of several epochs. Agassiz died before these men published. He thus missed the pioneer's greatest reward, that of seeing a new generation . . . refine his discoveries.

In Massachusetts, as in many other parts of this country and the world, the land surface has some special features. Among them are great boulders found resting on ledges of entirely different rock; sand and gravel found in tumbled piles; and exposed rock surfaces that appear smoothed as by a giant machine, with here and there grooves and scratches gouged across them. Such features have been explained as the work of great masses of ice on the land. These masses, called glaciers, build up where snow accumulates faster than it melts, and subsequently compacts into solid ice. Some glaciers have formed in mountains and moved down valleys; others have covered continents, as on Greenland and Antarctica today.

Only a little over a hundred years ago, the idea that glaciers might once have been much more extensive than they are today was novel and ridiculed. It had its modest beginnings in the Alps, where Louis Agassiz adopted and nourished it. But modern concepts of past ice ages were not fully developed until after his death in 1873.

When he was faced with the suggestion that glaciers had moved beyond their present limits, Agassiz proceeded logically by making measurements to learn the characteristics and habits of existing glaciers. He was surprised to find that the Aar glacier was nearer a thousand feet thick than the eighty to a hundred he had expected. He found that the Alpine glaciers were moving, even though their surfaces were melting and their snouts were not advancing, and that they were moving more in their middle portions than along the sides. Then he went on to observe how glaciers carry earth and rocks, how they abrade ledge surfaces with sand and boulders locked in their icy grip, how the things glaciers are doing today explain otherwise puzzling landscape features in regions far from existing icecaps.

The following article tells in Agassiz's own words how he pieced the evidence together, starting with the first year when he had 100 feet of iron rods lugged up the Aar glacier only to find that he might as well "have tried to sound the ocean with a 10-fathom line!" The article was published in 1865, but it is a timeless account of geologic observation and reasoning.

Louis Agassiz

ICE ON THE LAND

When I first began my investigations upon glaciers in the Swiss Alps, scarcely any measurements of their size or their motion had been made. One of my principal objects, therefore, was to measure them. I started by trying to ascertain the thickness of the mass of ice, generally supposed to be from 80 to 100 feet and even less. The first year I took with me 100 feet of iron rods (no easy matter, where it had to be transported to the upper part of a glacier on men's backs), thinking to bore the glacier through and through. As well might I have tried to sound the ocean with a 10-fathom line! The following year I took 200 feet of rods with me, and again I was foiled. Eventually, I succeeded in carrying up 1,000 feet of line, and satisfied myself after many attempts that this was about the average thickness of the glacier of the Aar, on which I was working.

A like disappointment awaited me in my first attempt to ascertain by direct measurement the rate of motion in this mountain glacier. Early observers had asserted that the glacier moved, but there had been no accurate demonstration of the fact, and so uniform is its general appearance from year to year that even the fact of its motion was denied by many.

My first experiment was to plant stakes in the ice to ascertain whether they would change their position with reference to the sides of the valley. If the glacier moved, my stakes must of course move with it; if it was stationary, my stakes would remain standing where I had placed them. I learned nothing from this experiment, except that the surface of the glacier is wasted annually for a depth of at least 5 feet, in consequence of which my rods had lost their support and fallen down.

My failure, however, taught me to sink the next set of stakes 18 feet below the surface of the ice, instead of 5; and the experiment was attended with happier results. A stake planted 18 feet deep in the ice, and cut on a level with the surface of the glacier in the summer of 1840, was found in September of 1841 to project 10 feet above the surface. Before leaving the glacier this time, I planted six stakes in a straight line across the upper part of the glacier, taking care to have the position of all the stakes determined with reference to certain points fixed on the rocky walls of the valley. When I returned the following year, all the stakes had advanced considerably, and the straight line had changed to a crescent, the central rods having moved forward much faster than those nearer the sides, so that not only was the advance of the glacier clearly demonstrated, but also the fact that its middle portion moved faster than its margins. This furnished the first accurate data concerning the average movement of the glacier during the greater part of one year. . . .

Glaciers also cover extensive areas beyond the limits of mountain valleys in Spitzbergen and Greenland. These Arctic lands are famous for their extensive ice sheets which come down to the seashore, where huge masses of ice, many hundreds of feet in thickness, break off and float away into the ocean as icebergs. Icebergs were first traced back to their true origin by the nature of the land ice of which they are always composed, which is quite distinct in structure and consistency from the marine ice produced by frozen sea water, and called "ice flow" by the Arctic explorers.

Sea, pond, and river ice result from the simple freezing of water at a certain temperature (under the same law of crystallization by which any inorganic body in a fluid state may assume

ICE ON THE LAND

a solid condition). It is easily recognized by its stratification, with beds varying in thickness according to the duration of the cold.

Land ice, of which both ice sheets and mountain glaciers consist, is produced by the slow and gradual transformation of snow into ice. Water readily penetrates the porous snow and sinks by its own weight; in time the whole mass becomes more or less filled with moisture. When the temperature falls below freezing, the water becomes frozen and fills the snow with ice particles. By this process, a mass of snow may be changed to a kind of granular ice-gravel. In time, the frost will transform the crystalline snow into more or less compact ice composed of an infinite number of aggregate snow particles of irregular shape and cemented together by the ice formed from the infiltrated moisture, the whole being interspersed with air.

The lower portions of a glacier are forced out by the pressure of the superincumbent ice, producing movement in the glacial ice. When confined by rock barriers as in a mountain valley, and on an inclined plane, the movement is mostly downward; but in an ice sheet many thousands of feet thick, covering usually all but the highest mountains in its path, the motion is outward from the region of great snowfall.

Land ice, when exposed to a temperature sufficiently high to dissolve it, does not melt from the surface and disappear by a gradual diminution of its bulk like pond ice, but crumbles into its original granular fragments, each one of which melts separately. This accounts for the sudden disappearance of icebergs, which instead of slowly dissolving into the ocean, are often seen to fall to pieces and vanish at once.

There is no chain of mountains in which the shape of the valleys is more favorable to the formation of glaciers than the Alps. These huge mountain troughs form admirable cradles for the snow, which collects in immense quantities within them, and, as it moves slowly down from the upper ranges, is transformed into ice on its way and compactly crowded into the narrower space below. At the lower extremity of the glacier, the ice is

Aletsch glacier, Swiss Alps. Lake Marjelen in the foreground is almost dry. (Harvard University, Department of Geology.)

pure, blue, and transparent, but, as we ascend, it appears less compact, more porous and granular, assuming gradually the character of snow, till in the higher regions the snow is as light, as shifting, and incoherent, as the sand of the desert.[1]

There are many mountain chains as high or higher than the Alps, which can boast of but few and small glaciers, if, indeed, they have any. In the Andes, the Rocky Mountains, the Pyrenees, the Caucasus, the few glaciers remaining are insignificant in size. The volcanic, conelike shape of the Andes gives but little chance for the formation of glaciers, though their summits are capped with snow. The height of these mountains is such that, were the shape of their valleys favorable to the accumulation of snow, they might present beautiful glaciers. In the Tyrol, on the contrary, as well as in Norway and Sweden, we find glaciers almost as fine as those of Switzerland, in mountain ranges much lower than either of the above-named chains. But they are of diversified forms, and have valleys widening upward on the slope of long crests.

There are certain facts which identify present glaciers and are proof of their former greater extension. First, there is the singular abrasion of the surfaces over which a glacier has moved, quite unlike that produced by the action of water. Such surfaces,

[1] A snowstorm on a mountain summit is very different from a snowstorm on the plain, on account of the different amounts of moisture in the atmosphere. At great heights, there is never dampness enough to allow the fine snow crystals to coalesce and form what are called snowflakes.

wherever the glacier marks have not been erased by some subsequent action, have several unfailing characteristics: they are highly polished, and they are also marked with scratches or fine *striae*, with grooves and deeper furrows. Where best preserved, the smooth surfaces are shining; they have a luster like stone or marble artificially polished by the combined friction and pressure of some harder material than itself until all its inequalities have been completely leveled and its surface has become glossy.

The leveling and abrading action of water on rock has an entirely different character. Tides or currents driven powerfully and constantly against a rocky shore, and bringing with them hard materials, may produce blunt, smooth surfaces, such as the repeated blows of a hammer on stone would cause; but they never bring it to a high polish, because the grinding materials are not held down in firm permanent contact with the rocky surfaces against which they move, as is the case with the glacier. On the contrary, being dashed to and fro, they strike and rebound, making a succession of blows, and never a continuous, uninterrupted pressure and friction. All the marks are separate, disconnected, fragmentary; whereas the lines drawn by the hard materials set in the glacier, whether light and fine or strong and deep, are

A glacially worn outcrop of rock at Glacier Creek, Montana, showing the characteristically smoothed, scratched, and grooved surface. (Northern Pacific Railway.)

continuous, often unbroken for long distances, and straight. Indeed, we have seen that we have beneath every glacier a complete apparatus adapted to all the results described above. In the softer fragments ground to the finest powder under the incumbent mass we have a polishing paste; in the hard materials set in that paste, whether pebbles, or angular rocky fragments of different sizes, or grains of sand, we have the various graving instruments by which the finer or coarser lines are drawn. Not only are these lines frequently uninterrupted for a distance of many yards, but they are for the most part parallel.

Certain inequalities of glacier-worn surfaces do exist, however, and deserve special notice. These are knolls, which seen from a distance resemble the rounded backs of a flock of sheep resting on the ground. They result from the prolonged abrasion of large masses of rock separated by deep indentations wide enough to be filled by large glaciers. The glacier moving as a solid mass, and carrying on its under side its gigantic file, will in course of time abrade uniformly the angles against which it strikes, equalizing the depressions between the prominent masses, and rounding them off until they present smooth bulging knolls. Besides their peculiar form, these knolls all have the characteristic features of glaciation in their polished surfaces accompanied with the straight lines, grooves, and furrows described above.

In enumerating the evidences of glacier action, we have to remember not only the effects produced upon the surface of the ground by the ice itself, but also the deposits it has left behind.

A steadily melting valley glacier drops its rocky fragments uniformly along the valley. It is where the glacier is still receiving enough new snow upstream to keep its ice in motion, but melting at its lower end is fast enough to keep the front from advancing for a period of time, that large deposits can accumulate.

These deposits of rock materials, such as boulders, gravel, sand, and clay, are the retreating footprints of a glacier, as it slowly yielded possession of the plain, and betook itself to the mountains; wherever we find one of these walls of unusual size, there we may be sure the glacier resolutely set its icy foot, disputing the ground inch by inch, while heat and cold strove for the mastery. There may have been a succession of cold summers, so that the glacier stayed for a number of years at the same line, and added constantly to the debris collected at its lower extremity. Wherever such pauses in the retreat of the glacier occurred, all the materials annually brought down to its terminus were collected; and when finally it disappeared from that point, it left a wall of debris, called a terminal moraine, to mark its temporary resting place.

Other moraines, called lateral moraines, mark the limits of the valley glaciers which once occupied crescent-shaped depressions. These lateral moraines often bar the outlet of waters from glaciers above, forming lakes.

Some moraines may be obliterated by meltwater from wasting ice; others may be simply remodeled by washing away of glacier mud and resulting rough sorting of the material. We have here a blending of the action of water with that of the glacier; and, indeed, how could it be otherwise, when the colossal glaciers of the past ages gradually disappeared or retreated to the mountain heights?

Moraines mark the retrogression of a glacier; for had these successive walls of loose materials been deposited in consequence of the advance of the glacier, they would have been pushed together in one heap at its lower end. They also indicate the relative

Map of North America showing the maximum extent of Pleistocene continental glaciers, indicated by curved lines. (W. C. Alden, U.S. Geological Survey.)

age of ancient moraines, not only by their position with reference to each other and to the present glacier, but also by their vegetation. The older ones have a mature vegetation and the more recent ones are almost bare of vegetation.

Greenland and the Arctic regions hold all that remains of the glacial ice sheet of northern America. Their shrunken ice fields, formidable as they seem to us, are to the frozen masses of that secular winter but as the patches of snow and ice lingering on the north side of our hills after the spring has opened. Let us expand them in imagination until they extend over half the continent, and we shall have a sufficiently vivid picture of this frozen world.

Let us for a moment assume that an accumulation of glacier ice takes place in the far north and acquires a thickness of from 12,000 to 15,000 feet. Such a mass would in consequence of its own weight uniformly advance in a southerly direction from the Arctic toward the more temperate latitudes in Europe, Asia, and North America. But we need not build up a theoretical case in order to form an approximate idea of the great ice sheet stretching over the northern part of this continent during the glacial period. It would seem that man was intended to decipher the past history of his home, for some remnants or traces of all its great events are left as a key to the whole.

Mountain ranges give us a means of measuring the thickness of the ice sheet which once covered America. If a mountain, for instance, is over 6,000 feet high, and the rough unpolished surface of its summit, covered with loose fragments, just below the level of which glacier marks come to an end, tells us that it lifted its head alone above the desolate waste of ice and snow, then, the thickness of the sheet cannot have been much less than 6,000 feet. . . . In North America, wherever the mountains are much below 6,000 feet, the ice seems to have passed directly over them, while the few peaks rising to that height are left untouched.

The glaciers which now descend through all the valleys of the Alps, once covered the plain of Switzerland and that of northern Italy.

The Scotch Highlands and the mountains of Wales and Ireland are but a few of the many centers of glacier distribution in Europe. From the Scandinavian mountains, glaciers descended also to the shores of the Northern Ocean and the Baltic Sea. There is not a fiord of the Norway shore that does not bear upon its sides the tracks of the great masses of ice which once forced their way through it, and thus found an outlet into the sea. Indeed, under the water, as far as it is possible to follow them through the transparent medium, I have noticed in Great Britain and in the United States the same traces of glacial action, so that these prior glaciers must have extended not only to the seashore, but into the ocean, as they do now in Greenland. Nor is this all. Scandinavian boulders, scattered upon English soil and over the plains of northern Germany, tell us that not only the Baltic Sea, but the German Ocean also, was bridged across by ice on which these masses of rock were transported. In short, over the whole of northern Europe, from the Arctic Ocean to the northern borders of its southern promontories, we find all the usual indications of glacial action, showing that a continuous sheet of ice once spread over nearly the whole continent, while from all mountain ranges descended those more limited glacial tracks terminating frequently in transverse moraines across the valleys, showing that, as the general ice sheet broke up and contracted into local glaciers, every cluster or chain of hills became a center of glacial dispersion, such as the Alps are now, such as the Jura, the Highlands of Scotland, and many others have been in ancient times.

Glacial ice represents a part of the world's total supply of water; it is derived by evaporation from the oceans and by precipitation as snow, then trapped as it turns to ice and remains on the land. This would become a factor of geologic significance if glaciers incorporated enough water to affect the level of the oceans.

At first the idea of removing from the vast ocean expanses enough water to lower sea level was startling. But as understanding of the spread and thickness of continental ice sheets grew, it became apparent that sea level had been lowered by the removal of water to form the ice sheets. Once this was realized, evidence from the land was recognized. Oceans cannot throw their waves and currents at shorelines for long without cutting terraces to wave base, building bars, and otherwise leaving traces of their presence.

Lowering of sea level is, of course, reversed when glaciers melt and their water returns to the oceans. In fact, if all the ice now on Greenland and Antarctica were to melt, sea level would be a great deal higher than it is now. Sea level has been higher in the past, and evidence of higher stands is easier to find than that of lower stands, whose shorelines are now under water.

Swings of sea level caused by the glacial and interglacial periods of the past have amounted to hundreds of feet at least. One lowering of the oceans left the bottom of the North Sea exposed to the air, with the Rhine and Thames flowing across it. Paleolithic man lived when glaciation was extensive, seas were low, and he could walk over lands now submerged. Many seashore camps of early man are probably covered today by the flooding waters returned by melting glaciers.

For an account of all this, we draw again upon the treasury of Rachel Carson's **The Sea Around Us.**

Rachel L. Carson

ICE AND OCEAN LEVELS

Four times in the past million years, ice caps formed and grew deep over the land, pressing southward into the valleys and over the plains. And four times the ice melted and shrank and withdrew from the lands it had covered. We live now in the latest stages of this fourth withdrawal. About half the ice formed in the last Pleistocene glaciation remains in the ice caps of Greenland and Antarctica and the scattered glaciers of certain mountains.

Each time the ice sheet thickened and expanded with the unmelted snows of winter after winter, its growth meant a corresponding lowering of the ocean level. For, directly or indirectly, the moisture that falls on the earth's surface as rain or snow has been withdrawn from the reservoir of the sea. Ordinarily, the withdrawal is a temporary one, the water being returned via the normal runoff of rain and melting snow. But in the glacial periods the summers were cool, and the snows of any winter did not melt entirely but were carried over to the succeeding winter, when the new snows found and covered them. So, little by little, the level of the sea dropped as the glaciers robbed it of its water, and the climax of each of the major glaciations left the ocean all over the world standing at a very low level.

Today, if you look in the right places, you will see the evidences of some of these old stands of the sea. Of course, the strand marks left by the extreme low levels are now deeply covered by water and may be discovered only indirectly by sounding. . . .

Out of the hard granite of the island of Torghatten, the pounding surf of a flooding interglacial sea cut a passageway through the island, a distance of about 530 feet, and in so doing removed nearly 5 million cubic feet of rock. The tunnel now stands 400 feet above the sea.

During the other half of the cycle, when the seas sank lower and lower as the glaciers grew in thickness, the world's shorelines were undergoing changes even more far-reaching and dramatic. Every river felt the effect of the lowering sea; its waters were speeded in their course to the ocean and given new strength for the deepening and cutting of its channel. Following the downward-moving shorelines, the rivers extended their courses over the drying sands and muds of what only recently had been the sloping sea bottom. Here the rushing torrents—swollen with melting glacier water—picked up great quantities of loose mud and sand and rolled into the sea as a turbid flood.

Torghatten Island, off the coast of Norway, testifies to an earlier stand of the sea at the level of the dotted line, where the cliffs were notched by wave action and a 530-foot-long tunnel was cut through the granite of the island. These features are now 400 feet above the sea. (C. A. Ericksen.)

During one or more of the Pleistocene lowerings of sea level, the floor of the North Sea was drained of its water and for a time became dry land. The rivers of northern Europe and of the British Isles followed the retreating waters seaward. Eventually, the Rhine captured the whole drainage system of the Thames. The Elbe and the Weser became one river. The Seine rolled through what is now the English Channel and cut itself a trough out across the continental shelf—perhaps the same drowned channel now discernible by soundings beyond Lands End.

The greatest of all Pleistocene glaciations came rather late in the period—probably only about 200 thousand years ago, and well within the time of man. The tremendous lowering of sea levels must have affected the life of Paleolithic man. Certainly he was able, at more than one period, to walk across a wide bridge at Bering Strait, which became dry land when the level of the ocean dropped below this shallow shelf. There were other land bridges, created in the same way. . . .

Many of the settlements of ancient man must have been located on the seacoast or near the great deltas of the rivers, and relics of his civilization may lie in caves long since covered by the rising ocean. Our meager knowledge of Paleolithic man might be increased by searching along these old drowned shorelines. One archaeologist has recommended searching shallow portions of the Adriatic Sea, with "submarine boats casting strong electric lights" or even with glass-bottomed boats and artificial light in the hope of discovering the outlines of shell heaps—the kitchen middens of the early men who once lived there. Professor R. A. Daly has pointed out:

> The last Glacial stage was the Reindeer Age of French history. Men then lived in the famous caves overlooking the channels of the French rivers, and hunted the reindeer which throve on the cool plains of France south of the ice border. The Late-Glacial rise of general sea level was necessarily accompanied by a rise of the river waters downstream. Hence, the lowest caves are likely to have been partly or wholly drowned. . . . There the search for more relics of Paleolithic man should be pursued.

Some of our Stone Age ancestors must have known the rigors of life near the glaciers. While men as well as plants and animals moved southward before the ice, some must have remained within sight and sound of the great frozen wall. To these, the world was a place of storm and blizzard, with bitter winds roaring down out of the blue mountain of ice that dominated the horizon and reached upward into gray skies, all filled with the roaring tumult of the advancing glacier, and with the thunder of moving tons of ice breaking away and plunging into the sea.

But those who lived half the earth away, on some sunny coast of the Indian Ocean, walked and hunted on dry land over which the sea, only recently, had rolled deeply. These men knew nothing of the distant glaciers, nor did they understand that they walked and hunted where they did because quantities of ocean water were frozen as ice and snow in a distant land.

In any imaginative reconstruction of the world of the Ice Age, we are plagued by one tantalizing uncertainty: how low did the ocean level fall during the period of greatest spread of the glaciers when unknown quantities of water were frozen in the ice? Was it only a moderate fall of 200 or 300 feet—a change paralleled many times in geological history in the ebb and flow of the epicontinental seas? Or was it a dramatic drawing down of the ocean by 2,000 or even 3,000 feet?

Each of these various levels has been suggested as an actual possibility by one or more geologists. Perhaps it is not surprising that there should be such radical disagreement. It has been only about a century since Louis Agassiz gave the world its first understanding of the moving mountains of ice and their dominating effect on the Pleistocene world. Since then, men in all parts of the earth have been patiently accumulating the facts and reconstructing the events of those four successive advances and retreats of the ice. Only the present generation of scientists, led by such daring thinkers as Daly, have understood that each thickening of the ice sheets meant a corresponding lowering of the ocean, and that with each retreat of the melting ice a returning flood of water raised the sea level.

Of this alternate robbery and restitution most geologists have taken a conservative view and said that the greatest lowering of the sea level could not have amounted to more than 400 feet, possibly only half as much. Most of those who argue that the drawing down was much greater base their reasoning upon the submarine canyons, those deep gorges cut in the continental slopes. The deeper canyons lie a mile or more below the present sea level. Geologists who maintain that at least the upper parts of the canyons were stream-cut say that the sea level must have fallen enough to permit this during the Pleistocene glaciation.

In May, 1960, a series of earthquakes devastated 600 miles of Chile's coast. The sixth and largest of the series, on May 22, did more, for it caused a shift of the sea bottom which started a train of waves in the ocean. However, these were no ordinary waves. As they raced across the Pacific at 425 miles an hour, they were no more than 3 feet high and 100 miles from crest to crest —harmless and unnoticed. But the whole great depths of the ocean's water moved with them and when the bottom shallowed as they approached a shore, the tremendous energy of movement was compressed into less and less depth until it could find release only by pushing up the surface water into a hurtling, relentless wall. Sixteen hours after the earthquake, the first giant wave swept onto Hawaiian shores. When the last was gone, 51 people were dead. But the waves raced on, and nine hours later dealt death and destruction along the coast of Japan.

Destruction was inevitable, but loss of life was greatly reduced and could have been avoided entirely if the endangered had heeded warnings issued before the first wave arrived. Warnings of such giant waves are made possible by instruments. The system is described in the following article, which includes also a full discussion of "tidal waves," correctly known as tsunami. The article "Tsunami" appeared in **Scientific American,** and was written by Joseph Bernstein, an oceanographer with the U.S. Navy Hydrographic Office.

Joseph Bernstein

GIANT WAVES

On the morning of April 1, 1946, residents of the Hawaiian Islands awoke to an astonishing scene. In the town of Hilo almost every house on the side of the main street facing Hilo Bay was smashed against the buildings on the other side. At the Wailuku River a steel span of the railroad bridge had been torn from its foundations and tossed 300 yards upstream. Heavy masses of coral, up to four feet wide, were strewn on the beaches. Enormous sections of rock, weighing several tons, had been wrenched from the bottom of the sea and thrown onto reefs. Houses were overturned, railroad tracks ripped from their roadbeds, coastal highways buried, beaches washed away. The waters off the islands were dotted with floating houses, debris and people. The catastrophe, stealing upon Hawaii suddenly and totally unexpectedly, cost the island 159 lives and $25 million in property damage.

Its cause was the phenomenon commonly known as a "tidal wave," though it has nothing to do with the tidal forces of the moon or sun. More than 2,000 miles from the Hawaiian Islands, somewhere in the Aleutians, the sea bottom had shifted. The disturbance had generated waves which moved swiftly but almost imperceptibly across the ocean and piled up with fantastic force on the Hawaiian coast.

Scientists have generally adopted the name "tsunami," from

the Japanese, for the misnamed tidal wave. It ranks among the most terrifying phenomena known to man and has been responsible for some of the worst disasters in human history. What made the 1946 tsunami especially notable was that a number of oceanographers happened to be in the Pacific (in connection with the Bikini atomic bomb test) and were able to observe it at first hand. It became one of the most thoroughly investigated tsunami in history, and from it came the development of an effective new warning system by the U.S. Coast and Geodetic Survey.

A tsunami may be started by a sea bottom slide, an earthquake or a volcanic eruption. The most infamous of all was launched by the explosion of the island of Krakatoa in 1883; it raced across the Pacific at 300 miles an hour, devastated the coasts of Java and Sumatra with waves 100 to 130 feet high, and pounded the shore as far away as San Francisco.

The ancient Greeks recorded several catastrophic inundations by huge waves. Whether or not Plato's tale of the lost continent of Atlantis is true, skeptics concede that the myth may have some foundation in a great tsunami of ancient times. Indeed, a tremendously destructive tsunami that arose in the Arabian Sea in 1945 has even revived the interest of geologists and archaeologists in the Biblical story of the Flood.

One of the most damaging tsunami on record followed the famous Lisbon earthquake of November 1, 1755; its waves persisted for a week and were felt as far away as the English coast. Tsunami are rare, however, in the Atlantic Ocean; they are far more common in the Pacific. Japan has had 15 destructive ones (eight of them disastrous) since 1596. The Hawaiian Islands are struck severely an average of once every 25 years.

In 1707 an earthquake in Japan generated waves so huge that they piled into the Inland Sea; one wave swamped more than 1,000 ships and boats in Osaka Bay. A tsunami in the Hawaiian Islands in 1869 washed away an entire town (Ponoluu), leaving only two forlorn trees standing where the community had been. In 1896 a Japanese tsunami killed 27,000 people and swept away 10,000 homes.

The dimensions of these waves dwarf all our usual standards of measurement. An ordinary sea wave is rarely more than a few hundred feet long from crest to crest—no longer than 320 feet in the Atlantic or 1,000 feet in the Pacific. But a tsunami often extends more than 100 miles and sometimes as much as 600 miles from crest to crest. While a wind wave never travels at more than 60 miles per hour, the velocity of a tsunami in the open sea must be reckoned in hundreds of miles per hour. The greater the depth of the water, the greater is the speed of the wave; Lagrange's law says that its velocity is equal to the square root of the product of the depth times the acceleration due to gravity. In the deep waters of the Pacific these waves reach a speed of 500 miles per hour.

Tsunami are so shallow in comparison with their length that in the open ocean they are hardly detectable. Their amplitude sometimes is as little as two feet from trough to crest. Usually it is only when they approach shallow water on the shore that they build up to their terrifying heights. On the fateful day in 1896 when the great waves approached Japan, fishermen at sea noticed no unusual swells. Not until they sailed home at the end of the day, through a sea strewn with bodies and the wreckage of houses, were they aware of what had happened. The seemingly quiet ocean had crashed a wall of water from 10 to 100 feet high upon beaches crowded with bathers, drowning thousands of them and flattening villages along the shore.

The giant waves are more dangerous on flat shores than on steep ones. They usually range from 20 to 60 feet in height, but when they pour into a V-shaped inlet or harbor they may rise to mountainous proportions.

Generally the first salvo of a tsunami is a rather sharp swell, not different enough from an ordinary wave to alarm casual observers. This is followed by a tremendous suck of water away from the shore as the first great trough arrives. Reefs are left high and dry, and the beaches are covered with stranded fish. At Hilo large numbers of people ran out to inspect the amazing spectacle of the denuded beach. Many of them paid for their curiosity with their lives, for some minutes later the first giant wave roared over the shore. After an earthquake in Japan in 1793 people on the coast at Tugaru were so terrified by the extraordinary ebbing of the sea that they scurried to higher ground. When a second quake came, they dashed back to the beach, fearing that they might be buried under landslides. Just as they reached the shore, the first huge wave crashed upon them.

A tsunami is not a single wave but a series. The waves are separated by intervals of 15 minutes to an hour or more (because of their great length), and this has often lulled people into thinking after the first great wave has crashed that it is all over. The waves may keep coming for many hours. Usually the third to the eighth waves in the series are the biggest.

Among the observers of the 1946 tsunami at Hilo was Francis P. Shepard of the Scripps Institution of Oceanography, one of the world's foremost marine geologists. He was able to make a detailed inspection of the waves. Their onrush and retreat, he reported, was accompanied by a great hissing, roaring and rattling. The third and fourth waves seemed to be the highest. On some of the islands' beaches the waves came in gently; they were steepest on the shores facing the direction of the seaquake from which the waves had come. In Hilo Bay they were from 21 to 26 feet high. The highest waves, 55 feet, occurred at Pololu Valley.

Scientists and fishermen have occasionally seen strange by-products of the phenomenon. During a 1933 tsunami in Japan the sea glowed brilliantly at night. The luminosity of the water is now believed to have been caused by the stimulation of vast numbers of the luminescent organism *Noctiluca miliaris* by the turbulence of the sea. Japanese fishermen have sometimes observed that sardines hauled up in their nets during a tsunami have enormously swollen stomachs; the fish have swallowed vast numbers of bottom-living diatoms, raised to the surface by the disturbance. The waves of a 1923 tsunami in Sagami Bay brought to the surface and battered to death huge numbers of fishes that normally live at a depth of 3,000 feet. Gratified fishermen hauled them in by the thousands.

The tsunami-warning system developed since the 1946 disaster in Hawaii relies mainly on a simple and ingenious instrument devised by Commander C. K. Green of the Coast and Geodetic Survey staff. It consists of a series of pipes and a pressure-measuring chamber which record the rise and fall of the water surface. Ordinary water tides are disregarded. But when waves with a period of between 10 and 40 minutes begin to roll over the ocean, they set in motion a corresponding oscillation in a column of mercury which closes an electric circuit. This in turn sets off an alarm, notifying the observers at the station that a tsunami is in progress. Such equipment has been installed at Hilo, Midway, Attu and Dutch Harbor. The moment the alarm goes off, information is immediately forwarded to Honolulu, which is the center of the warning system.

This center also receives prompt reports on earthquakes from four Coast Survey stations in the Pacific which are equipped with seismographs. Its staff makes a preliminary determination of the epicenter of the quake and alerts tide stations near the epicenter for a tsunami. By means of charts showing wave-travel times and depths in the ocean at various locations, it is possible to estimate the rate of approach and probable time of arrival at Hawaii of a tsunami getting under way at any spot in the Pacific. The civil and military authorities are then advised of the danger, and they issue warnings and take all necessary protective steps. All of these activities are geared to a top-priority communication system, and practice tests have been held to assure that everything will work smoothly.

Since the 1946 disaster there have been 15 tsunami in the Pacific, but only one was of any consequence. On November 4, 1952, an earthquake occurred under the sea off the Kamchatka Peninsula. At 17:07 that afternoon (Greenwich time) the shock was recorded by the seismograph alarm in Honolulu. The warning system immediately went into action. Within about an hour, with the help of reports from seismic stations in Alaska, Arizona and California, the quake's epicenter was placed at 51 degrees North latitude and 158 degrees East longitude. While accounts

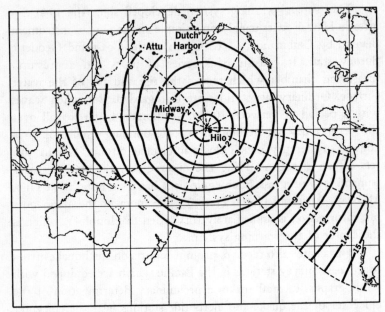

Estimated time required for tsunami to reach the Hawaiian Islands from earthquakes in different parts of the Pacific. The numbers represent hours. For example, if a tsunami were generated by an earthquake located just south of the Aleutian Islands, it should reach the Hawaiian Islands about four and a half hours after the earthquake occurred. Seismic sea-wave detectors are operated at Attu, Dutch Harbor, Hilo, and Midway. (After Joseph Bernstein in Scientific American, *August, 1954.)*

of the progress of the tsunami came in from various points in the Pacific (Midway reported it was covered with nine feet of water), the Hawaiian station made its calculations and notified the military services and the police that the first big wave would arrive at Honolulu at 23:30 Greenwich time.

It turned out that the waves were not so high as in 1946. They hurled a cement barge against a freighter in Honolulu Harbor, knocked down telephone lines, marooned automobiles, flooded lawns, killed six cows. But not a single human life was lost, and property damage in the Hawaiian Islands did not exceed

$800,000. There is little doubt that the warning system saved lives and reduced the damage.

But it is plain that a warning system, however efficient, is not enough. In the vulnerable areas of the Pacific there should be restrictions against building homes on exposed coasts, or at least a requirement that they be either raised off the ground or anchored strongly against waves.

The key to the world of geology is change; nothing remains the same. Life has evolved from simple combinations of molecules in the sea to complex combinations in man. The land, too, is changing, and earthquakes are daily reminders of this. Earthquakes result when movements in the earth twist rocks until they break. Sometimes this is accompanied by visible shifts of the ground surface; often the shifts cannot be seen, but they are there; and everywhere can be found scars of earlier breaks once deeply buried. Today's earthquakes are most numerous in belts where the earth's restlessness is presently concentrated, but scars of the past show that there is no part of the earth that has not had them.

The effects of earthquakes on civilization have been widely publicized, even overemphasized. The role of an earthquake in starting the destruction of whole cities is tremendously frightening, but fire may actually be the principal agent in a particular disaster. Superstition has often blended with fact to color reports.

We have learned from earthquakes much of what we now know about the earth's interior, for they send waves through the earth which emerge with information about the materials through which they have traveled. These waves have shown that 1,800 miles below the surface a liquid core begins, and that it, in turn, has a solid inner core.

Earthquakes originate as far as 400 miles below the surface, but they do not occur at greater depths. Two unsolved mysteries are based on these facts. (1) As far down as 400 miles below the surface the material should be hot enough to be plastic and adjust itself to twisting forces by sluggish flow rather than by breaking, as rigid surface rocks do. (2) If earthquakes do occur at such depths, why not deeper?

Knowledge gained from studying earthquake waves has been applied in various fields. In the search for oil and gas, we make similar waves under controlled conditions with dynamite and learn from them where there are buried rock structures favorable to the accumulation of these resources. We have also developed techniques for recognizing and locating underground nuclear tests through the waves in the ground which they generate.

The following discussion of this subject has been adapted from the book **Causes of Catastrophe** by L. Don Leet.

L. Don Leet

THE RESTLESS EARTH AND ITS INTERIOR

At twelve minutes after five on the morning of Wednesday, April 18, 1906, San Francisco was shaken by a severe earthquake. A sharp tremor was followed by a jerky roll. The roar and crash of man-made structures mingled with a dull booming from the earth itself. Within about a minute, the shaking tapered off and the disturbance ceased.

A number of buildings were shaken into various stages of collapse. The new 7 million dollar City Hall was one of these. Many frame buildings were wrenched and distorted, some by the slumping of loose soil under their foundations, but the damage to well-designed and solidly built structures was less than was first thought. The 14-year-old 11-story Crocker Building, the steel-framed James Flood Building, the Mint, the new Post Office Building, the partially complete Hotel St. Francis, and scores of others escaped lightly. Authoritative estimates charge the earthquake's shaking with responsibility for less than 5 per cent of the total damage to property on that day. This was what is known as the San Francisco earthquake, one of history's most important.

Before people had gained the open air and begun to take

stock of what had happened, puffs of smoke were mingling with the great cloud of dust that rose from the city. Realization of the full sweep of disaster came slowly. The day wore on and people discussed their experiences. Some mourned their dead. All paused in momentary panic as repeated aftershocks swept under the city and on to the eastern horizon, but none of these matched the first great tremor, and gradually the tenseness of the populace eased. Then the significance of a growing pall of smoke over Market Street and a widening area began to appear. Pressure in the city water mains, which had been broken in many places, was insufficient for combating the conflagration, and the fire got out of control. Residents of the threatened districts began a trek to open spaces in outlying sections. Three days passed. Then, by dynamiting structures in their path, flames were robbed of fuel. Finally, rain fell. Meanwhile crowds of refugees were huddled together without elementary sanitary conditions, and to cap a week's nightmare, were attacked by epidemics of filthborne disease. This was the San Francisco fire, which occasioned property losses estimated at 400 million dollars.

Deaths from this earthquake along the California coast probably did not exceed 700, approximately that occasioned by the New England hurricane of September 21, 1938, or by automobiles in the United States in almost any week.

Many earthquakes elsewhere have exceeded this one in intensity and caused incomparably greater loss of life and property. Actually, this earthquake shook a few hundred miles of California's coast much more severely than San Francisco that first morning, for in reality the earthquake did not occur at San Francisco at all. And its historical importance does not rise from the fact that the new 7 million dollar City Hall and many other buildings were ruined or that a spectacular and disastrous fire followed. It stands as one of the world's most important earthquakes because of the striking nature of the clues it left as to its cause and because of the thorough scientific investigation based on it.

Traversing California's coastal ranges of mountains for some

600 miles is a scar in the earth's crust, which geologists call a fault. It has been called the San Andreas Rift or San Andreas Fault. It is one of thousands, if not millions, which mark the globe's surface and extend downward for many miles.

Around 1800 all was peaceful and at rest along the northern part of the San Andreas Rift. Then, urged by internal forces which are constantly at work, the bordering land was gradually shoved and twisted until by 1875 the region along the fault had become considerably warped. The shoving and bending continued until, on the morning of April 18, 1906, the rocks were strained to the limit of their strength. At 5:12 A.M., they broke along a 200-mile section of the fault and snapped back into a nearly unstrained position. The resulting horizontal displacements were as great as 21 feet in some places, and the vibrations set up by the violent adjustment traveled outward in all directions. At San Francisco, 10 miles away from the nearest part of the fault, they were strong enough to be felt and to do damage. They coursed on through and around the earth, gradually dying out, but with enough energy to leave records on seismographs, instruments designed to record such disturbances, on the other side of the world.

Fences and roads which crossed the fault were offset. A walk leading up to the steps of a ranch house near Olema hopped northwestward several feet, while the house was jerked in the opposite direction. Water pipes crossing the fault were disrupted.

Such extensive surface displacements are rare phenomena. In no other earthquake on record has an equal length of fault displacement been traced. However, there are instances of vertical uplift of portions of the earth's crust. Along the coast not far from Tokyo, a record of such a rise during historic time is to be found engraved in stone. In that region lives a certain boring bivalve known as *Lithophaga nasuta*, which occupies a cigar-shaped pair of shells a few inches in length and drills a home into the rocky shore at mean sea level. He lives on organisms brought to his door by the sea. At one place sets of bore holes have been found at four different levels. A bit of deduction based

Rock Creek slide across the Madison River, Montana, caused by the Yellowstone earthquake of August 17, 1959. (Gordon Oakeshott.)

on historical records has attributed their present positions above sea level to elevations of the land which occurred at the times of earthquakes in the years 33, 818, 1703, and 1923. The total uplift was 45 feet.

In 1899, near Yakutat Bay, Alaska, a series of major shocks occurred. Investigators later found remains of recently dead sea animals on a wave-cut platform 50 feet above sea level. This elevation had been accomplished in a single jump, putting Alaska three jumps ahead of Japan.

The San Andreas Rift displacements, change of level of land near Tokyo and in Alaska, and other similar cases supply significant evidence of the mechanism involved in snapping of the earth's crust to cause earthquakes. They represent only a small percentage of the world's earthquakes, however; the great majority leave no such obvious clues at the surface.

Earthquakes are caused by breaks in solid rocks of the earth, sometimes far below the surface, sometimes near or at the surface. These breaks invariably result from pressures which shove great blocks about, but always work to push them against each other. The pressures cannot, and never do, pull them apart to leave yawning chasms in the solid crust. Sometimes, however, the shaking that results when crustal breaks occur jars masses of unconsolidated materials so that they slump or slide into new positions. The natural cracks and fissures that result in such deposits are no more fearful or remarkable than they would have been had the slump or slide occurred after a hard rain. These indirect effects of earthquakes are really normal, familiar processes.

The point at which the break begins, called the focus, is ordinarily from 10 to 30 miles deep. Vibrations from such a focus are called normal, or shallow-focus earthquakes and constitute from 80 to 90 per cent of the recorded shocks. Scattered through the records of modern seismograph stations, however, are a few which differed radically from the general run of these normal earthquakes. Just a few years ago, an explanation for these was found, and they can now be interpreted. They come from foci which are deeper than normal. The deepest which have been observed to date were 435 miles below the surface of the earth, over a tenth of the way to the center. Maybe that does not sound like much, but scientifically this was an exciting discovery because before it had been made there was believed to be no doubt, on theoretical grounds, that rocks at such depths could not have enough strength to accumulate stresses which, when suddenly relieved, would cause earthquakes.

The present distribution of deep-focus earthquakes also is unique. One region in which they are fairly common is South America. There the deepest ones occur inland, shallower ones toward the coast, and the normal shocks nearest the coast and offshore. A depth of 400 miles is favored by the South American variety. Japan is another scene of deep-focus earthquakes. There

A 20-foot vertical scarp in limestone up Red Canyon about 2 miles northeast of Hebgen Lake, Montana. This was formed by displacements of the surface by faulting at the time of the Yellowstone earthquake of August 17, 1959. (Gordon Oakeshott.)

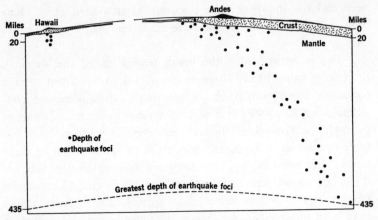

Earthquake foci under Hawaii and South America.

the favored depth is nearer 220 miles. Deep-focus earthquakes are relatively numerous in the South Pacific Ocean. An isolated group of 140 miles depth is located near the Pamir Plateau in the Hindu Kush Mountains of the Himalayan region.

Deep-focus earthquakes are bringing to light many new and interesting details of our globe's restless surface. They appear to be outlining for us the important structural trends of some of the youngest mountains of this geologic era and the deep-seated movements controlling present volcanic activity.

When an earthquake occurs, it sets up vibrations in the ground. Sound is a vibration in the air which produces a familiar effect on our eardrums. Sound is alternating compressions and rarefactions of the air, which travel outward from a source as waves. If these occur 435 times a second, we hear a note called middle A. If they are irregular and jumbled, we hear a noise which may be low and booming or high and shrill, depending on the proportions of slow or rapid compressions in the jumble.

During an earthquake, there are sometimes vibrations in the ground which disturb the air in such a way as to produce sounds within the range of the human ear's receiving band. These are earthquake sounds. They have been variously described, and are usually placed on the low booming side of the scale. Very near the source of an earthquake, the sounds sometimes include sharp snaps and suggest the tearing apart of great blocks of rocks. Farther away, they have been likened to the sound of heavy vehicles passing rapidly over hard ground or a road; dragging of heavy boxes or furniture over the floor; a loud but distant clap of thunder; an explosion, or the boom of a distant cannon; the fall of heavy bodies, or loads of stone or coal falling.

The vibrations in the earth which constitute the shaking we call an earthquake are not simple things. They are in the form of waves, by which energy is transmitted through and around the earth. These waves are classified as push, shake, and surface waves.

When a break in the earth's rocks occurs, both push and shake waves are generated simultaneously. They start out in all directions from their point of origin but at different speeds. It is as though a streamlined passenger train and a fast freight started together on parallel tracks. After several miles away, one will be definitely ahead of the other. Of the earthquake waves, the push wave is the faster. It reaches distant points before the shake wave, and accordingly was originally called simply the primary wave. By the same token, the wave coming in second was called the secondary wave. From these names came the symbols which are commonly used for them, P and S. The surface waves are the largest and longest waves, and are thought to be generated when P and S cause large displacements at the surface near their place of origin. They travel at the slowest speed of all, a mere 100 miles a minute. The symbol L is used for surface waves, originally because of their largeness. Surface waves sometimes stretch out 100 miles or more from crest to crest. Their height depends in part upon the size of P and S near the focus, where surface waves are generated. Accordingly, they will be very small or entirely absent in the case of a deep-focus earthquake, when P and S have to travel a relatively long distance before reaching the surface to generate surface waves.

The difference between speeds of P and S is a great convenience to seismologists. After the waves start simultaneously from an earthquake's focus, P gains progressively on S. At a place 100 miles away, P arrives 20 seconds before S gets there; 1,000 miles away, P is 2 minutes and 40 seconds ahead of S and 4 minutes ahead of L; 2,000 miles away, P leads S by 4 minutes and 52 seconds, and at 7,175 miles by an even 12 minutes, with L at this last distance more than half an hour behind P.

Seismographs show when the ground is disturbed by vibrations of any of these types. They make their records with time marks so that the exact time of any disturbance, as well as its existence, is recorded. Accordingly, when an earthquake like that at San Francisco in 1906 has a place of origin which is fairly well known from surface evidence or distribution of its surface effects, and the distances from that to various seismograph stations that recorded the vibrations are thus known, it is possible to make time tables for the intervals between P, S, and L as they reach known distances. Also, by working backward, it is possible to figure the instant at which P and S were together, that is, the time of the earthquake at its source. When this has been done, the intervals between waves can be supplemented by the exact length of time required for them to reach the different distances.

Tables such as Tables I and II are essential tools of the seismologist. When the records of a station, which are ordinarily changed daily, show an earthquake with P, S, and L clearly evident, the observer determines the intervals between them. By using the interval table, he can find at once the distance to which the intervals apply. For example, if he observes that S arrived

Record of waves 5,300 miles from an earthquake. (Harvard University's Oak Ridge Seismograph Station.)

8 minutes after P, he concludes from the table that the earthquake was 4,000 miles away. He then notes that P arrived at, say, 12 minutes and 22 seconds after 4:00 A.M. The travel-time table, shows that P requires 9 minutes and 50 seconds to travel 4,000 miles. Accordingly, it started that length of time before it reached the station in question, so the time of the earthquake at its source was 4:12:22 minus 9 minutes 50 seconds, or 4:02:32; that is, 2 minutes and 32 seconds after 4:00 A.M.

TABLE I   INTERVAL TABLE FOR P, S, AND L

Distance from source, miles	Time interval between P and S (S minus P)		Time interval between P and L (L minus P)	
	Min	Sec	Min	Sec
100		20		
1,000	2	45	4	00
2,000	4	52	8	20
3,000	6	30	13	30
4,000	8	00	18	00
5,000	9	25	24	30
6,000	10	44	29	00
7,000	11	49	35	20

TABLE II   TRAVEL-TIME TABLE FOR P, S, AND L

Distance from source, miles	Travel time for P		Travel time for S		Travel time for L	
	Min	Sec	Min	Sec	Min	Sec
100		27		47		
1,000	3	20	6	00	7	20
2,000	5	56	10	48	14	16
3,000	8	00	14	30	21	30
4,000	9	50	17	50	27	50
5,000	11	26	20	51	35	56
6,000	12	43	23	27	41	43
7,000	13	50	25	39	49	10

This process is used by all seismograph stations which record a quake. They should, of course, be in close agreement, regarding the time they compute for the instant of an earthquake. Arcs drawn on a globe about each station, with radii equal to the computed distances, intersect at or near the center of the disturbance, and the earthquake is located.

In addition to traveling with different speeds, P and S differ in another important respect. The P wave can travel in a solid, liquid, or gas, but the S wave can exist only in materials which resist attempts to shear, that is, distort them or change their shape. Such materials do not include liquids and gases, molecules of which move over and around each other with the greatest ease. This difference has an important bearing on seismological information about the earth's interior.

P, S, and L are not the only waves resulting from an earthquake. As the waves from an earthquake are traced outward at successively more distant seismograph stations, it is found that something happens just beyond 7,000 miles from the earthquake's source. The first recorded wave fails to conform to the schedule established up to that point for P; it is late. Worst of all, S disappears, or gets so weak that it is lost. This does not mean that the record is a blank until surface waves arrive. There are several scores of combination waves which result from P and S bouncing around inside the earth. These combination waves change from one type to another and back again, and in a variety of ways complicate the pattern. A distant seismograph station records the arrival of various kinds of waves for hours after a large earthquake.

When we say that S disappears, or gets lost, we mean that at the time set by schedules for shorter distances, there is at distances beyond 7,000 miles no longer a prominent wave of the type S had been up to then. Nor does such a wave appear a short time later, as it would if S were merely delayed as P was. S is not there at all.

It can be shown that P waves which reach a surface point about 7,000 miles from an earthquake have had to penetrate

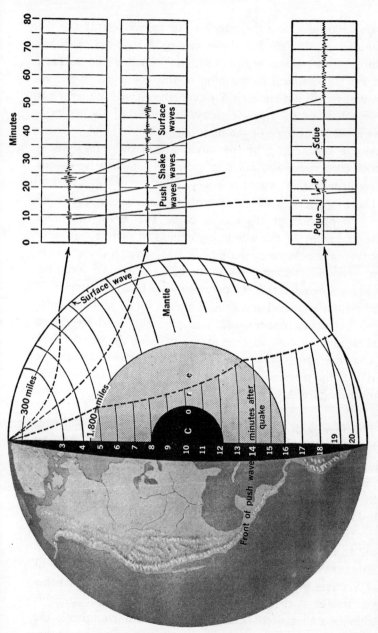

Internal structure of the earth, with lines representing successive positions of the first wave from an earthquake each minute after the quake occurred. Dotted lines represent the paths of travel from the source to surface points at distances of 4,800 miles, 6,900 miles, and 9,900 miles from the earthquake.

close to 1,800 miles into the interior. Hence, we learn from the strange case of the late P and the missing S that the earth has a definite core, which begins 1,800 miles below the surface. Whatever the chemical constitution of this core, it is in a condition which will not transmit S waves; that is, it is either a gas or a liquid or of such low rigidity that most of the shake is taken out of S and some of the push out of P.

At the very center of the earth, within the fluid core, is an inner core about 800 miles in radius in which wave speeds increase again as they would if it were solid.

There is evidence which controls to some extent guesses as to the composition of the earth's core. The earth, the other planets, and many of the wandering chunks of matter that enter our atmosphere as meteorites are believed to have had a common origin. Accepting this premise, in the best tradition of Aristotle, it could be concluded that the core of the earth is predominantly a combination of nickel and iron, which is found in large numbers of meteorites. Astronomers and physicists lend weight to this view in their computation of the mass of the earth. They find that the earth cannot be as light throughout as are the materials at the surface, if their computations are correct—which they probably are. A heavy core would take care of the situation.

The first push waves start away from the focus much as would ripples from a pebble dropped into a quiet pond, and reach the opposite side of the globe in 20 minutes and 7 seconds. En route, when they strike the core, they are bent as light waves are by a lens. As a consequence certain distances cannot be reached either by the direct P wave or by the one which has been bent by the core and has emerged on the far side. These distances are said to be in a shadow zone for the direct push waves.

From studies of the paths, speeds, and general behavior of P and S waves, seismologists have concluded that the earth has a crystalline crust that is about 20 miles thick under continents and thinner under deep ocean basins. Beneath continents, the crust is dominantly granitic in composition, grading into a ba-

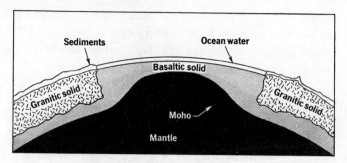

The earth's crust and mantle under continents and oceans.

saltic layer which also underlies the deep oceans. At the base of the crust, the rocks change abruptly into a new material which possesses unique physical properties.[1] These include the capacity to yield by slow plastic flow to changing loads and squeezing pressures during mountain making. Yet once mountains are formed, it is dense enough to "float" them. Also, this same material is brittle enough to break to cause earthquakes. When we consider all this, we are faced with the fact that these are an unusual set of apparently contradictory properties. Our expanding experience with the behavior of matter, however, has shown that contradictory or not, these properties can be possessed by a single substance, as they are, for instance, by silicone putty, which is dispensed as "silly putty" for children and geologists to play with.

Another change occurs at 300 miles; then there seem to be no sharp boundaries until the core is reached at 1,800 miles.

At the present time, instrumental records are being used to locate an average of from 600 to 700 earthquakes a year. Of these, from 50 to 100 may be literally world shakers capable of causing catastrophe if they occur near a large center of population. If we include everything down to the smallest trucklike tremors

---

[1] The surface that separates the crust from the new material (called the mantle) is known as the Moho discontinuity; Moho is an abbreviation of the name of Mohorovicic, the first seismologist to report this surface.

unrecorded in remote spots, following major earthquakes by the thousands as aftershocks, it is likely that a figure of 50,000 earthquakes a year—more than 150 a day—is short of the true number. One seismologist's estimate reached 1,000,000 a year.

The majority of earthquakes tend to cluster in two zones. One of these borders the Pacific Ocean. The other starts in the Mediterranean basin and Central Europe, and swings east through Asia and southeast along the Malay Peninsula, joining the first zone in the South Pacific. A minor zone follows the buried ridge along the axis of the Atlantic Ocean. About 75 per cent of the world's earthquakes are now occurring in these zones.

Historical records show that a given region may undergo a spasm of activity followed by a lull of two or more centuries. Since our modern instrumental records cover only a quarter of a century, it is inevitable that there are many seismic sections of the globe not represented in current statistics. There are probably not many, if any, square miles of the earth's surface which have not experienced and will not at some future date again experience earthquake vibrations strong enough to be felt.

Earthquakes, like so many things in nature, occur in cycles, but not with sufficient regularity to permit even a rough guess as to when another one is due. Tokyo, during the past 250 years, has had several severe earthquakes. There was a major one on December 31, 1703. One of almost equal severity occurred on October 28, 1707. Others were felt: 64 years later in April, 1771; 84 years later on November 11, 1855; and 68 years later on September 1, 1923. The 1703 and 1923 shocks were from practically the same source.

Northeastern America, though not in the principal zones of present activity, has a history of seismicity for which records of the past 300 years have been kept. They include a major earthquake at Boston in 1755. Investigators have found evidence of a progressive tilting of the ground resulting in subsidence of the land relative to sea level at the rate of a foot per century at Boston and about two-thirds that amount at New York. The restlessness of the crust necessary for the production of earthquakes appears to be present here.

Nine centuries are covered by records from Lisbon, Portugal, which start in 1009, culminate in one of the world's greatest known earthquakes on November 1, 1755, and carry through to today. The intensity of the earliest shocks is in doubt, but a major shock in 1344 was followed 187 years later by a principal shock in 1531. This, in turn, was followed 224 years later by the shock of 1755. Each principal shock was preceded by foreshocks and followed by aftershocks. Some of the most severe were entered in the record. Aftershocks following the earthquake of 1755 continued until 1858, since which time the region has been quiescent. Thus, Lisbon and Boston, which had their most recent major earthquakes in the same month nearly two centuries ago, are good candidates for a return to the active class within this century.

Volcanoes stir the imagination, often to a point where wildly inaccurate concepts of them are formed. In 1902, Mount Pelée on the island of Martinique wiped out the city of Saint-Pierre with its thousands of inhabitants; and the fate of the cities of Pompeii and Herculaneum when Vesuvius erupted in 79 A.D. is well known. As a result of ideas about volcanic eruptions based on such incidents, during World War II it was proposed that hostilities in Italy be ended by bombing Vesuvius to trigger an eruption, and in Japan by bombing Mihara, on the island of Oshima just outside Tokyo Bay. These were unrealistic proposals, however, because if a volcano is not primed by internal forces and ready to erupt, bombing it would be like trying to fire an unloaded gun by pulling the trigger. Moreover, many if not most eruptions are merely scenic attractions for neighbors and visitors.

Such an eruption occurred in 1959 at Kilauea on the island of Hawaii where warnings of an impending eruption began months before any lava appeared at the surface. Instrumental measurements showed that the volcano was swelling from increasing internal pressures as lava moved into position several miles below the surface. Then the lava seemed to stop pressing upward; it even appeared to withdraw somewhat for several months; but it resumed its outward pressure on the volcano and finally, to the accompaniment of thousands of small earthquakes, burst forth.

The following selection is adapted from "The 1959 Eruption of Kilauea" by Jerry P. Eaton and Donald H. Richter, published in **Geotimes.** The authors are scientists of the Hawaiian Volcano Observatory who witnessed this eruption and who, moreover, have participated in the program of scientific studies that are giving us the best clues yet developed as to eruptive mechanisms.

Jerry P. Eaton and Donald H. Richter

A VOLCANO ERUPTS

"There it goes!" "In Kilauea Iki!" "I saw it first!" shrilled voices in the crowd outside the Observatory as a bright orange fume cloud suddenly lit the night sky over Kilauea Iki Crater, 2½ miles east of the U.S. Geological Survey's Hawaiian Volcano Observatory.

This was the dramatic beginning of the 1959 eruption of Kilauea, at 8:08 in the evening of November 14.

Kilauea is a broad basaltic shield volcano on the island of Hawaii, southernmost of our newest state's island chain. The summit caldera, with its inner pit, Halemaumau, the usual site of Kilauea summit eruptions, is within the boundaries of Hawaii National Park. Kilauea Iki, or little Kilauea, scene of the 1959 eruption, is a pit crater about a mile long by half a mile wide that is separated from the main summit caldera by a low, narrow ridge called Byron Ledge.

Although the eruption appeared at the surface with electrifying suddenness, it had its real beginning quietly, months before and miles beneath the outbreak point. The first of a network of liquid-level tiltmeter bases was installed near the observatory in November, 1957, and in the following months it indicated the ground surface at that station to be steadily inclining outward from the caldera. As we brought additional tilt bases around the caldera under measurement in the months that followed, the pattern became clearer: the whole caldera region was bulging upward and tilting outward. Analysis of tilting between October, 1958, and February, 1959, suggested that magma was welling up quietly from the depths and accumulating steadily in a zone several miles beneath the south rim of the caldera.

Following several moderate earthquakes just southeast of the caldera on February 19, 1959, the swelling stopped, and from May until August the summit of the volcano subsided slowly. This we interpreted as a recession of magma in the column. Then a great swarm of earthquakes and associated tremor (over 2,500 quakes and many hours of tremor) between August 14 and 19 originating about 35 miles beneath the north rim of Kilauea caldera were recorded by the U.S. Geological Survey's seismograph net on Hawaii. Magma moving into the deep volcanic plumbing system during this episode made itself felt at the surface shortly, for rapid swelling of Kilauea resumed between August and October.

In mid-September a very sensitive seismograph at the northeast edge of Halemaumau, near the center of the caldera, began recording a swarm of tiny quakes originating less than half a mile away. Although these quakes were exceedingly small, their number was impressive: over 22,000 by November 14. Except for their smaller size, these quakes closely resembled those preceding the 1955 eruption of Kilauea from its east rift zone. As the number of quakes increased, scientific and public interest began to mount. Uncertain of the significance of these tiny shallow quakes, we began a hurried remeasurement of tilt-base changes in early November. Dramatic changes had occurred: the caldera region was swelling at least three times faster than had been detected before. On the evening of November 14 as the observatory staff readied its special eruption equipment "just in case" and as an excited crowd of "seismograph watchers" gathered outside the observatory window, the tiltmeter crew started off to remeasure the last two tilt bases; but the eruption broke out before the crew got to work.

View of lava fountain in Kilauea Iki; the fountain rose 600 to 900 feet. Taken from Byron's Ledge at 6 A.M. on November 19, 1959. The level of lava in the lake is approaching that of the opening from which the fountain comes. (U.S. Geological Survey.)

Minutes after the first glow was sighted, we were standing on the rim of Kilauea Iki with one of nature's most fantastic displays unrolling before us. Starting in a single fissure half way up the 600-foot south wall of the crater, fountaining rapidly spread laterally in both directions. By 10 P.M. ten short fissures, each with one or more active lava fountains, formed a discontinuous "curtain of fire" ½ mile long. A few minutes after this maximum lateral development, the fountaining at the ends of the line of fissures began to abate. Gradually activity ceased in the outermost vents and by 4 A.M. only two fountains remained. One of these continued through the early afternoon of November 15; then it too died. The other grew in size, eventually reaching the unprecedented height of 1,900 feet on December 17 during the fifteenth phase of the eruption.

During the first night and early morning we began to suspect, as the line of fountains shrank, that the eruption might be over by dawn. In view of the possibility of a short-lived eruption, we were anxious to start collecting gas and lava samples. The best sampling area appeared to be at the base of the largest fountain, where small lava flows occasionally branched off the main stream and covered a few hundred square feet before solidifying. Clad in asbestos protective gear we descended into the crater and walked over still warm flows and vents that had died only a few hours earlier. We got within 200 feet of the base of the fountain. At the time, however, it seemed more like 50 feet than 200; with brilliant orange-yellow spatter being tossed into the air above our heads, horizontal distances were foreshortened drastically. With the first 1959 samples collected, we hastily retreated to the crater wall, to a firmer and much cooler vantage point above the fountain. Little did we realize that only two weeks later the area where we had stood collecting the samples would be under 100 feet of lava and have a 1,000-foot fountain playing above it.

From the scientific point of view, we have been very fortunate to have the eruption confined to a pit crater practically in our own backyard. Not only has this allowed easy access for sampling and observation, but we have also been able to calculate rather precisely the volume and rate of lava extrusion and withdrawal much the same as in a graduated cylinder. From the human viewpoint, however, we have mixed feelings with regard to the proximity of the eruption. Throughout the course of the eruption the fountain has been very unpredictable, sometimes low and bubbling, but more often shooting high in sudden repeated bursts, depositing pumice as far as 5 miles to the southwest. Had the wind shifted during any of the eruptive periods, and blown from the south, as it often does in Hawaii during the winter season, the housing area for the National Park would have received a blanket of basaltic pumice up to 2 feet thick. Persistent trade winds have also spared us from the acrid and nauseating gas fumes which persist for as many as 50 miles downwind from the vent.

As of December 17, 1959, just over one month after the initial outbreak, and the date of part of this writing, we have had a total of 15 separate eruptive phases. During the week-long first phase 40 million cubic yards of lava were poured into

Jerry Eaton (foreground) and Wayne Ault sampling hot gases during the 1959 eruption of the Hawaiian volcano, Kilauea Iki. (U.S. Geological Survey.)

Lava pond at the extreme eastern end of Kilauea Iki on December 15, 1959, showing collapse of the central portion of the pond due to withdrawal of fluid lava following the eleventh eruptive phase. (U.S. Geological Survey.)

Kilauea Iki, forming a lava pond 335 feet deep. The 14 subsequent eruptions have been of much shorter duration and have contributed an additional 17 million cubic yards of lava to the pond, increasing its depth to a maximum of 414 feet at the end of the eighth phase. Although the duration of the later eruptive phases has decreased, the rate of lava output has increased. In the first stage a maximum of 470,000 cubic yards per hour was measured during the last day of that phase; the tenth phase spewed out its lava at the phenomenal rate of 1,400,000 cubic yards per hour!

The temperature of the lava in the fountain measured consistently above 2050°F. On December 5, 2200°F was recorded, which as far as we can determine is the highest temperature ever observed for Hawaiian volcanoes. The high temperatures seem to correlate with the generally "primitive" undifferentiated nature of the lava being erupted.

The large pumice and cinder cone which is being built in the lee of the fountain, had an approximate elevation of 3,910 feet at the end of the seventh phase of activity, or about 150 to 200 feet above the irregular platform on which it rests. Chester K. Wentworth, geologist, is investigating the physical properties of the pyroclastic debris which forms the cone and is following the changes in volume of erupted material. The cone is already a dominant physiographic feature in the National Park.

One of the many interesting facets of the 1959 eruption is the withdrawal of liquid lava back into the vent when fountaining ceases. This phenomenon has been noticed ever since the end of the second phase when the level of the lava in the pond rose above the volcanic vent. In fact from the end of the sixth phase almost all the lava erupted has poured back down the vent. As viewed from above along the crater's rim, the backward flow of lava toward and down the hole has at times resembled a bathtub full of water with its drain plug removed. Even the rings show as the lava drains out and the pond surface sinks. The rates for this withdrawal are also phenomenally high. Although less accurately determined than the extrusive rates, backflow has been measured at rates exceeding 2 million cubic yards per hour or almost twice the average rate of extrusion. Whether the lava that

Geophysicist H. L. Krivoy inspecting campground shelter, denuded ohia forest, and 4-foot deposit of basaltic pumice 3,500 feet southwest of the Kilauea Iki eruption, on December 15, 1959. (U.S. Geological Survey.)

has been extruded during the later stages represents an entirely fresh batch or merely recycled pond lava with enough primary magma added to recharge it with gas for the push to the surface, is a significant problem that is not yet resolved. Jack Murata, who has been turning out silica analyses almost as fast as the lava is being ejected, remarks that it would be a shame if he were analyzing the same old stuff over and over again. Per cent silica varied between 46.3 and 49.5 during the early phases but has more or less stabilized at 46.8 since the fourth phase. We are beginning to think in terms of "lava geysers" at the observatory.

We hope that this very informal story of the 1959 Kilauea eruption will serve to answer a few of the questions that undoubedly have arisen since the eruption began and to show briefly what investigations are being made by the U.S. Geological Survey. Needless to say the writing of this article has been rather hectic. We started at the end of the seventh eruptive phase, but were interrupted by phases eight through fifteen. We hope that before too long we can go back to more normal hours—whoops, have to wait a little while longer, phase sixteen has just started!

Since the earth's surface first solidified, volcanoes have been adding gases to the atmosphere and water to the earth's supply. Rock-making materials have poured out of them in vast quantities, and volcanic products are found in rocks of all ages on all parts of the globe. Modern volcanoes provide natural laboratories for the study of the processes that are involved in all this.

No two volcanoes are exactly alike in their habits or histories. From some, hot lavas pour out quietly and spread over wide areas. From others, material is blown out violently in a series of explosions. Sometimes a single volcano will behave first in one then in the other of these fashions. Some volcanoes are quiescent for long intervals, even for hundreds of years, while others have been active almost continuously since far back in historic time.

In spite of their individuality, volcanoes are believed to share a common cause and certain general features. The original magma that supplies them all is probably similar at the depths from which it starts toward the surface. Differences develop along the route of rise, introduced by partial crystallization or by ingestion of rocks. All are actuated by gas pressures, though some have more than others. All have a finite life span, though for some it is much longer than for others. All, during their active time, appear to be located along zones where the earth's crust is buckling and contorting to form mountains, and breaking to cause earthquakes.

The story of the ultimate cause of volcanic action is told by Prof. Howel Williams, a world authority in this field and professor of geology at the University of California, in the following selection adapted from his monograph **The Ancient Volcanoes of Oregon.**

Howel Williams

VOLCANOES BUILD THE LAND

When we deal with the ultimate causes of volcanic action, we move in a field only dimly lit, stumbling in shadows of doubt. But even though speculation is rife, some generalizations appear to be valid. Records from deep wells and mines show that the temperature of the earth's crust increases with depth. The rate of increase varies. In the outer shells of the earth it averages between 86 and 122°F per mile; at greater depths the rate diminishes. Forty miles below the earth's surface the temperature is probably close to 2200°F. At that temperature in the laboratory (at atmospheric pressure), almost all rocks melt. Yet earthquakes demonstrate beyond question that the material 40 miles beneath us is not melted. What keeps it solid is the tremendous pressure of the overlying rock.

Twenty to forty miles under the floors of the oceans and under the continents, according to most geologists, there is an earth shell made up of heavy material of basaltic composition that grades downward into layers of still heavier rock material increasingly charged with nickel and iron. Some maintain that this subcrustal shell of basaltic material grades downward into a rock called peridotite, composed largely of the mineral olivine; others say that it passes downward into a metamorphic rock called eclogite; almost all agree that from time to time some of these materials are partly converted to a pasty liquid called magma, and that this is the primary source of all the lavas and ashes erupted by volcanoes.

How are the rigid rocks made liquid? The answer is: by reduction of pressure, by increase of temperature, or by a combination of these two processes. Now, as in the past, most volcanoes are concentrated in long, narrow belts across the face of the earth. Examples are the volcanoes of the Andes, the Cascades, the Aleutians, Japan, and the East Indies, parts of the "girdle of fire" encircling the Pacific Ocean.

These volcanic belts coincide closely with the major earthquake belts of the earth, for volcanoes and earthquakes alike are symptoms of unrest in the crust of the globe. In addition, most volcanoes lie within or close to mountain ranges that, by geologic standards, are youthful. The inference has therefore been drawn that bending and fracture of the earth's crust, by causing a local release of load at depth, convert some of the underlying hot rocks into magma.

Liquefaction may be brought about also by rise in temperature consequent to breakdown of radioactive substances in and beneath the crust. The elements uranium and thorium are especially important in this regard; as they decay to lead by giving off helium, they generate heat, and in the course of millions of years this may accumulate until large volumes of subcrustal material are changed to liquid.

Once magma is produced, it tends to rise. If ascent to the surface is rapid, the magma pours out of swarms of narrow fissures and spreads as floods of basaltic lava. Alternatively, the lava flows pile up to build giant basaltic volcanoes like those now active on the island of Hawaii. Usually, however, the rising magma is stopped temporarily at various levels in the earth's crust. Displacing the surrounding rocks, it comes to occupy reservoirs at depths of a few miles beneath the surface. These are the feeding chambers of most volcanoes.

It is well known that a single volcano may erupt quite different kinds of material at various times, and that neighboring

Schematic sketch of the island of Hawaii to illustrate that it is composed of five volcanoes which have been built up from the sea floor and have merged to form the island. This is the highest volcanic pile on earth, towering nearly 6 miles above the surrounding sea bottom. The volcanoes are (from background to foreground) Kohala, long extinct; Mauna Kea (on right), highest of the five and with no historical record of eruption; Hualalai (low cone on left), which last erupted in 1801; Mauna Loa, second highest and active at infrequent intervals; Kilauea, active about 66 per cent of the time from 1832 to 1955. (After Edward A. Schmitz, in L. D. Leet and S. Judson, Physical Geology, Prentice-Hall, Inc., Englewood Cliffs, N.J., 1958.)

volcanoes may discharge different lavas simultaneously. The explanation seems to be that the magma in the feeding chambers is always undergoing change, always tending to separate into fractions of different composition. As the liquid cools against its rocky walls, minerals begin to crystallize. Those forming first are usually poor in silica and rich in magnesia. As cooling proceeds, minerals richer in silica, iron, soda, and potash develop. Many of these crystals sink toward the bottom of the reservoir with the result that a light silica-rich residual liquid with few crystals comes to rest on a heavier liquid increasingly loaded with crystals toward the bottom. Since eruptions may take place at any stage in the process and eruptive fissures may tap any level in a feeding chamber, a wide variety of materials may be expelled.

If eruptions recur at brief intervals and the chamber is continually replenished from below, then the lavas and ashes are likely to be made up of olivine basalt, almost identical with the original magma. On the other hand, quiet intervals between eruptions may be long. Then, crystallization may continue until a light, siliceous liquid with sporadic crystals of quartz and associated minerals collects at the top of the reservoir. Underneath is a layer of intermediate composition devoid of quartz that rests in turn on heavy basaltic magma loaded with more and more iron-magnesian crystals toward the base. When layering of the reservoir progresses to this stage, the topmost quartz-bearing magma may be erupted to form rocks known as rhyolite or dacite. If lower layers of quartz-free magma of intermediate composition are erupted, the material hardens to rock referred to as andesite, the dominant product of the volcanoes crowning the Andes of South America. If still lower layers escape from the reservoir, they produce basaltic andesite and olivine basalt.

The process of crystallization is probably the main cause of the diversity in composition of the products of volcanoes. Among many other causes is the contamination of magma by solution of the reservoir walls. Any kind of rock may enclose a reservoir; hence the effects of solution in modifying the magma are extremely varied.

A fundamental effect of the crystallization of magma, the concentration of gas in the liquid that remains, must be emphasized. The reason is simple enough: none of the early-forming crystals abstracts gas from the magma. Consequently, the residual liquid becomes increasingly charged with volatile ingredients. Indeed, if crystallization goes on long enough, so much gas is concentrated in the remaining liquid that it can no longer be held in solution. Bubbles then begin to form; the magma starts to effervesce. Ultimately, the gas pressure becomes too great for the reservoir roof to withstand, and the frothy magma blasts a passage to the surface, exploding violently into showers of ash and pumice. None can doubt that this accumulation of gas pressure during crystallization is one of the prime causes of volcanic eruptions. And quite apart from the effects of crystallization, gas tends to diffuse and concentrate toward the upper and cooler parts of a magma chamber. Without gas, magma would be inert;

in large measure, it is the expansion of gas that forces magma upward to the surface and propels ejecta from the crater of a volcano.

The products of volcanoes include gases, lavas, and fragmental ejecta. Consider first the gases. By far the principal gas given off by volcanoes is steam or water vapor. Seldom does it constitute less than 80 per cent of the total discharge, and generally it makes up more than 95 per cent. Next in importance is carbon dioxide; then various compounds of sulfur, such as hydrogen sulfide and sulfur dioxide. Along with these, there is usually some carbon monoxide, hydrochloric and hydrofluoric acid, hydrogen, hydrocarbons, ammonium chloride, ammonia, etc. Even during a single eruption, the proportions of these minor constituents vary considerably. Their importance should not be minimized, however. Were it not for the emanations of volcanoes in the past, there might not be enough carbon dioxide in the atmosphere to support plant life, and without plants, man and animals could not exist.

Consider next the fragmental products of volcanoes. These range in size from blocks weighing hundreds of tons to particles fine enough to be carried by winds around the world. The finest ejecta, particles smaller than peas, are referred to as volcanic dust and ashes. Compacted to rocks, they form volcanic tuffs. Pieces between the size of peas and walnuts are spoken of as lapilli. Still larger fragments are called blocks, if already solid when blown out, or bombs, if partly or wholly in a molten state when erupted. Rocks consisting mainly of blocks are classified as volcanic breccias, while those mainly composed of bombs are termed agglomerates. Highly inflated, frothy ejecta, light enough to float on water, are designated as pumice; they are usually composed of dacite or rhyolite. Darker, clinkerlike lumps hurled out by basaltic volcanoes are commonly called cinders. Many small cones are built entirely of cinders.

The lavas erupted by volcanoes are no less diverse than the fragmental products. Their characters are controlled, likewise,

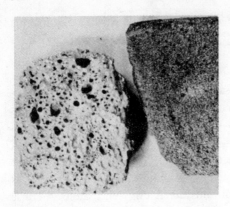

Pumice. Each specimen is approximately 4 inches long.

by the chemical composition, gas content, and temperature of the magma. Other things being equal, lavas poor in silica and rich in lime, iron, and magnesia, such as basalts, are more fluid than lavas like dacite and rhyolite in which the proportions of these constituents are reversed. Indeed, the most fluid basalts may pour along at the speed of a mountain stream, while rhyolitic and dacitic lavas crawl forward sluggishly. Hence it is not surprising that basaltic flows are usually much more extensive than siliceous ones. Besides, basaltic lavas are generally much hotter, their temperatures ranging mostly between 1800 and 2200°F, while rhyolitic and dacitic lavas vary normally between 1100 and 1550°F. Andesitic lavas tend to have intermediate temperatures. Cooler and more viscous rhyolitic and dacitic flows, therefore, form thick and stumpy tongues or steep-sided mounds, and they often solidify to the volcanic glass known as obsidian.

No one can travel through areas of recent volcanic activity without noting that the surface forms of the lavas are extremely diverse. Some flows, especially basaltic ones, have smooth, satiny skins of glass that glisten in the sunlight. Others have crusts marked by ropy and cordlike corrugations of the kind known in Hawaii as *pahoehoe*. It is in such flows that tubes and tunnels are best developed. Some of these tunnels are too small to crawl into; others measure 50 to 80 feet across and can be traced for a mile or more. Their origin is easy to understand. Lava solidifies

VOLCANOES BUILD THE LAND

first at the top, bottom, and sides, so that the interior continues to flow long after the marginal parts have come to rest. Hence, when the supply of fresh lava is checked or cut off at the source, the liquid interior may be drained by discharge at the snout of the flow, leaving the solid casing behind. The ceilings of many such tunnels are lined with slender stalactites caused by dripping of lava remelted by hot gases rising from the moving currents below. If the ceilings are thin, they may collapse to produce pits and elongate depressions on the surface of the flow.

Still other lavas, usually andesitic and basaltic ones, have

Ropy lava. (U.S. Geological Survey.)

indescribably rough, clinkery crusts that resemble seas of frozen foam. These the Hawaiians call *aa* flows. Then there are lavas having surfaces littered with chaotic piles of angular, smooth-faced blocks. Block lavas of this kind are typical of glassy, siliceous flows provided they chill quickly to form a thick crust of glass that can be shattered by movement of the pasty liquid underneath. Finally, some flows, particularly basaltic ones erupted into water, break up into pillow- and sack-shaped bodies. Excellent examples of such pillow lavas may be seen among the oldest volcanic rocks of the Coast Ranges of Oregon.

Perhaps the most familiar volcanic form is the graceful cone whose sides steepen toward the summit. Mounts Shasta, Hood, Rainier, and Saint Helens are splendid illustrations. Volcanoes like these are built partly of lava flows and partly of fragmental layers. In other words, they grow by a combination of quiet, effusive eruptions and violent explosions. Hence, they are commonly classed as composite volcanoes. When they rise to great height, the lavas tend to escape more and more from cracks far down the sides instead of from the crater at the top, although explosive blasts may continue from the summit and plugs of viscous lava may be forced upward through the crater floor.

Contrasted with composite cones are the so-called shield

Mount Hood, picturesque active volcano in Oregon. (Gardner Collection, Harvard University.)

volcanoes, built almost wholly by copious outwellings of fluid basalt. If the shields grow by overflows from a central vent on top and from more or less radial cracks on the flanks, they assume the forms of inverted saucers. If, on the other hand, overflows from the summit alternate with eruptions of lava from closely spaced, parallel fissures on the sides, the shields take on the shapes of inverted canoes. No better examples can be found of volcanic shields than Kilauea and Mauna Loa on the island of Hawaii. Many similar shields grew along the crest of the Cascade Range within the last ten million years.

Eruptions of clinkery ash, lapilli, and bombs produce the well-known cinder cones, such as Wizard Island in Crater Lake, Pilot Butte near Bend, and the scores of dark hillocks on the slopes of the Newberry Volcano in central Oregon. Few exceed 500 feet in height. The way in which they grow is exemplified by the activity of Parícutin in Mexico.

Not uncommonly, lava emerging from a vent is too viscous to spread far and therefore accumulates as steep-sided, bulbous mounds. Because such mounds are often of domical shape and serve to seal the underlying conduits, they are referred to as plug domes. Some grow by overflow of pasty lava from a crater on top; others, including Lassen Peak in California and some of the obsidian mounds in the Three Sisters region, Oregon, are forced from the feeding pipes much in the same way as toothpaste is squeezed from a tube. The outside of the lava column solidifies at once to form a glassy crust; then, as the pasty liquid within continues to rise, the crust is shattered into blocks that accumulate on the summit of the growing mass or tumble down the flanks to form long banks of talus. Compared with the rate of growth of composite and shield volcanoes, the rise of plug domes is phenomenally rapid. Volcanoes such as Mounts Hood, Rainier, and Shasta may have been a million years in the making; Lassen Peak and similar huge domes may have grown in less than a decade.

The activity of most volcanoes changes from time to time. Periods of violent explosions may alternate with periods of quiet effusion. Many vents may be active simultaneously within a single crater, each behaving in a different fashion. Despite these variations, certain well-known volcanoes erupt in a characteristic way for long periods; their names have thus come to be used in classifying types of eruption.

The Hawaiian type is exemplified by basaltic shield volcanoes like Mauna Loa and Kilauea. Extremely hot and fluid lavas pour from vents on the summits of the shields and from long fissures on the flanks. Sometimes fountains of lava spout during the first phases of an eruption, but the fragmental material blown out is trivial in volume compared with the lava flows.

The Strombolian type takes its name from the Italian volcano, Stromboli, which has been almost continuously active since the days of Homer. Normally, the mode of eruption is a more or less rhythmic discharge, every few seconds or minutes, of pasty, glowing clots of magma that cool to ropy, spindle- and almond-shaped bombs and clinkery lapilli. Quiet intervals are rarely long enough to allow lava to congeal in the feeding pipe; hence, few solid fragments are expelled. Outpouring of lava is on a much smaller scale than on Hawaiian volcanoes, and the flows are usually much more viscous. The characteristic form produced by Strombolian activity is a cinder cone.

Not far from Stromboli is Vulcano, from which the word "volcano" is derived and the Vulcanian type of eruption takes its name. Activity here is marked by discharge of still more viscous magma. Explosions, instead of being rhythmic and fairly continuous, take place between irregular intervals of repose. Solid, angular fragments are blown out, along with lumps of pasty magma that fall to earth as bombs with glassy crusts and as frothy pieces of pumice. Few fragments are hot enough to glow or liquid enough to be rounded as they spin through the air. Huge cauliflower clouds of steam, heavily charged with fine ash and riddled with flashes of lightning, rise from the crater. Flows are rare and those that do escape cool to thick, stumpy tongues of obsidian. Eruptions of this kind are exceptional on basaltic volcanoes; they are characteristic of volcanoes fed by more siliceous magmas.

When no lava is discharged during an eruption and the fragmental ejecta are made up entirely of old rock fragments, the activity is said to be of Ultra-Vulcanian type. Eruptions of this character are simply low-temperature steam blasts. The first outbreak of a new volcano and the initial explosions of volcanoes that have lain dormant for a long time are frequently of this type.

Ruins of the city of Saint-Pierre, Martinique, shortly after its destruction by a glowing avalanche from Mount Pelée on the morning of May 8, 1902. (Underwood and Underwood.)

In 1902, viscous lava was forced upward into the summit crater of the West Indian volcano, Mount Pelée. Unable to spread laterally, it piled over the vent as a bulbous dome. Similar domical protrusions are said to be of Pelean type. Often their rise is accompanied by explosions of frightful intensity. While the dome of Mount Pelée grew, repeated blasts of superheated steam shot from its sides, carrying with them vast quantities of glowing ash and blocks. So voluminous were these ejecta that they fell at once on the adjacent slopes, then raced down the mountain sides at hurricane speeds. Some of these glowing avalanches were observed to move at rates of more than 100 miles an hour. One overwhelmed the town of Saint Pierre in an instant, killing all but one of its 28,000 inhabitants.

Escape of lava from fissures on the sides of volcanoes rather than from central vents is a common phenomenon. But the most copious fissure eruptions are not those related to cones and shield volcanoes. On the contrary, they produce plains and plateaus of enormous extent. Many times during the earth's history, colossal floods of fluid basalt have risen through narrow, vertical fissures to spread over the surface in far-reaching floods, converting mountainous regions into level wastes. No less than a quarter of a million square miles of Oregon and Washington were inundated in this fashion, about 15 million years ago.

Although basaltic lava is the principal product of such large-scale fissure eruptions, the most siliceous magma, rhyolite, also may be poured out in immense volumes from narrow cracks in the ground. Usually, however, rhyolite is not erupted as flows, but as fragmental pumice and ash. Instead of being hurled high into the air, as in most explosive eruptions, the effervescing magma wells from the fissures as a mixture of hot gases, spray, and pasty clots. Having unusual mobility, the material spreads swiftly as incandescent sheets and travels far even over surfaces that are practically horizontal. About 35 million years ago, glowing avalanches of ash and pumice buried 80,000 square miles of western Utah and eastern Nevada, in places to a depth of 8,000 feet,

Giant's Causeway, near Portrush, Antrim, Northern Ireland, where basalt poured out onto the surface in a series of flows, each 10 to 50 feet thick, and hardened in a unique pattern of columnar joints. Each step on the skyline in the left background is the end of a flow. (David M. Owen.)

though the ancient surface across which the avalanches raced was almost flat.

From what has been said, it may be judged that the nature of volcanic eruptions is determined mainly by the gas pressure and viscosity of the magma involved. Other things being equal, the lower the viscosity the greater the tendency to quiet outflow of lava; the higher the gas pressure the greater the tendency to explosive activity. A magma with strong gas pressure may cause violent explosions; the same magma impoverished in gas may be forced out slowly to form a plug dome. The hotter a lava is and the more gas it contains, the more fluid it becomes. Composition is also important, for siliceous lavas are generally more viscous than basaltic ones. It is the complex interplay of all these and other factors which accounts for the multitude of ways in which volcanoes behave.

The history of Parícutin, the volcano born in Mexico in 1943, is of special interest; its activity shows how scores of cinder cones elsewhere have been formed.

The birth of a new volcano and the revival of activity of volcanoes after periods of repose are usually heralded by earthquakes. The birth of Parícutin was not an exception. For three weeks before the first eruptions, the ground in the vicinity shook almost continuously, and as the fatal day approached, the quakes increased in strength. On the morning of February 20, when Dionisio Pulido, a Tarascan Indian, went to till his corn patch he was amazed to see a wisp of vapor spiraling upward from a hole in the ground, a few inches wide. Within a few hours the wisp changed to a dark ash-laden column and the hole widened to 30 feet across. Late that night, glowing bombs and cindery clots began to issue, falling round the vent to build a cone. Next morning, the cone was already 120 feet high. Every few seconds, deafening blasts vomited showers of incandescent fragments, adding to its size. On the second day, a small tongue of basaltic lava emerged. It was amazing how rapidly the cone gained in height. On the third day, it was 200 feet high; on the twelfth, it was 450 feet high and lava had covered 120 acres. Enormous clouds towered over the summit of the volcano; sometimes they were shaped like a column that mushroomed at the top and sometimes like quickly expanding cauliflowers. They rose for 3 miles or more before being drifted away by the winds. Bombs up to several feet across rained down on the cone and around its base; farther away a steady shower of fine ash fell, laying waste the countryside. At night, the view was indescribably grand. Volleys of glowing projectiles, like fiery bouquets, shot from the crater. The cone sparkled with myriad moving lights, as bright red and golden bombs rolled and bounced down the sides while the overhanging clouds reflected a fitful, lurid glare. Above the deep roar of the cannonade could be heard the patter and thudding of falling

Parícutin, the volcano that broke out in a cornfield on February 20, 1943. The photograph shows the cinder cone as it appeared a few months later. The cone had reached a height of 1,200 feet nine years later, when the volcano died. (Charles B. Valla.)

fragments. The streams of lava flowing from the foot of the cone looked like incandescent ribbons. Without cease the ground trembled.

The noises of the explosions varied. For long spells, loud detonations recurred at intervals of a few seconds; then the sound changed to a dull, continuous roar like that of surf beating on a distant shore. The appearance of the eruption cloud also changed, passing from fleecy white to almost black as the amount of ash increased.

In the middle of March, following a period of exceptional violence, a new flow escaped from the base of the volcano. Early in April, activity became so intense that fine dust fell on Mexico City, 200 miles away. At Uruapan, 20 miles distant, the streets and housetops were heavily blanketed with cindery fragments. In mid-April a third flow issued from the foot of the cone. In June a fourth broke out from a point about halfway up the side. It undermined and carried away a large part of the cone. By mid-July the volcano was 1,000 feet high and more than 3,000 feet wide at the bottom. During September still another flow emerged from the foot of the cone. Throughout these months, explosions continued with unabated fury.

On October 19, 1943, a strange thing happened. Coincident with a sharp decrease in the explosive activity of Parícutin, a new cone, Zapicho, was born at its base. Spectacular reddish and golden yellow fountains of lava gushed from its mouth, and a long flow poured through a breach in its wall. For 79 days, until January 6, 1944, while Parícutin itself lay almost dormant, Zapicho erupted with vigor, building to a height of more than 200 feet. No sooner did it stop than Parícutin took up the refrain. Two flows burst from vents low on the flanks, and explosions from the summit-crater became so strong that heavy falls of ash were noted 100 miles away.

All through the early months of 1944, lava continued to pour from the base of the volcano. In May the principal flow, having traveled 5 miles, began to bury the town of San Juan; by late July, all but a small part of the town had been overwhelmed. Other flows burst from the base of the cone during the next few months. The village of Parícutin had long been rendered uninhabitable by heavy falls of ash. In October lava descended from the cone in a series of magnificent cascades and buried most of the buildings that remained. Early in November still another vent opened at the foot of the cone, and for more than three months lava issued from it in a steady stream, no matter whether the summit crater lay quiet or erupted with violence for days on end.

When the volcano was two years old, in February, 1945, the principal cone had a deep, funnel-shaped crater, approximately ¼ mile wide at the rim. On the crater floor were two vertical pipes that were astonishingly small considering the great volume of ash and bombs they discharged. Lavas had covered more than 4½ square miles, and close to the cone they had accumulated to a depth of about 600 feet.

After the second anniversary, there was little change in the type of activity. Quiet spells, when scarcely a wisp of vapor rose from the top of the cone, alternated with periods of strong explosions, sometimes as intense as those of the first few months. Flow after flow emerged from the foot of the cone; some lasted only for a few days, and others for a few weeks or months. Shortly after one ceased, another broke out from a nearby vent. The lava changed gradually from olivine-rich basalt or basaltic andesite to olivine-poor andesite, considerably richer in silica. At the same time, both the volume and rate of lava discharge diminished, though irregularly, and the successive flows tended to be shorter and more viscous. Close to the points of emission, their temperature was usually between 2100 and 2200°F, and even 2 or 3 miles from the vents, the central portions of some flows showed temperatures as high as 2000°F. Near the vents, most of the flows moved at rates varying from 3 to about 50 feet a minute, depending largely on the slope of the ground; at their snouts they crept forward only 60 to 600 or 900 feet a day.

The outstanding features of Parícutin's life were the phenomenal rate at which the cone grew and the great volume of lava expelled. Among the scores of flows erupted, all but a few broke out from vents at the base of the cone. Noteworthy also was the fact that while some flows were preceded by periods of strong explosions most of them behaved without apparent regard to the activity of the summit crater.

In March, 1952, when little more than nine years old, Parícutin died. The cone had reached a height of more than 1,200 feet above the corn field at its base. Altogether, some 2,500 million tons of ash and 1,500 million tons of lava had been erupted during this brief span of time.

The land is subject to continuous attack: rocks and minerals are broken by frost action and dissolved by acidic water, then the products of disintegration are moved about by water, wind, and ice. The earth strives to keep ahead by renewing itself with volcanic action and the action of other internal forces of unknown origin that fold and elevate sea-bottom rocks into high mountains. If the forces of building did not hold the upper hand, there would be no land, for there is enough water in the oceans to cover the entire globe to a depth of 2 miles if the rock surface were leveled.

The wearing away of the land can be seen in the billions of tons of solid and dissolved material carried to the sea each year by rivers. Sometimes in the gullying of a dirt slope by a hard rain, the effect of erosion on the land is clear. But more often the results are not plainly evident, even through many lifetimes. The Grand Canyon was obviously carved from solid rocks by the Colorado River and its tributaries, but there has been no significant change in it since it was discovered. To most persons, the highest peaks of the Himalayas would appear to be unchanging on the scale of human lifetimes, although a closer look at the base of each reveals piles of pieces broken from them. Changes occur slowly, but they are inevitable and continuous.

Wearing away of the land, the geologic process called erosion, is described by a famed British geologist, Prof. S. J. Shand, in the following selection adapted from his book **Earth Lore.**

S. J. Shand

WEARING AWAY OF THE LAND

The face of the earth, just like the human face, is always changing. Hills are no more "everlasting," in spite of what poets say, than flesh and bone, though they change more slowly. Every hill is draped with an apron of its own waste products, made up of great boulders, small stones, sand, and dust. On gentle slopes, this apron may be so thick that the massive rock is hidden beneath it. If the process of decay went on long enough, and the waste were not washed away or blown away, the hills would eventually be buried beneath their own ruins and the valleys would be filled up. That such an uninteresting climax is seldom reached, and then only in limited areas, we owe to the work of two tireless agents, water and wind.

A river, according to the dictionaries, is "a large running stream of water"; but this is only part of the story, because rivers carry a great deal more than water. It is impressive to be told that the Mississippi discharges into the Gulf of Mexico a million tons of water every minute; but it is far more interesting to a geologist to know that the river discharges in the same time 200 tons of dissolved lime and salts, and nearly 800 tons of silt. To carry and discharge such a load of mud, sand, and gravel would be utterly beyond the capacity of any railway system in the world. Professor Salisbury has made the impressive comparison that "it

Waste products of rock disintegration draping the slopes of mountains at Moraine Lake, Banff National Park, Alberta. (Gardner Collection, Harvard University.)

would take nearly nine hundred daily trains of fifty cars each, every car loaded with twenty-five tons, to carry an equal amount of sand and mud to the Gulf." The whole amount of dissolved and suspended matter brought down by the Mississippi in the course of a year comes to the huge total of 500 million tons; and it is judged that all the rivers in the world, taken together, discharge some 20,000 million tons of solids into the sea every year.

There is no mystery about the source of this material; it is the wastage of the lands. Wherever rocks are heated by the sun, wet by rain, and chilled by frost, they crack and break down. The process is slow, for nature seldom hurries, but not too slow for us to see it in action. The stone of many buildings and monuments, not necessarily of great age, has become cracked and discolored and has developed a crumbly, scaly surface, in the few score years since the building was put up. Such effects are caused partly by rapid changes of temperature, setting up strains between the grains of the rock, and partly by the chemical action of air and water. The net result of both processes is to break all exposed ock surfaces down to a mass of boulders, sand, and clay which, when it has become mixed with decaying vegetable matter, is what we call soil. Rain water soaking through the soil takes up whatever is soluble and carries it at once away to the rivers. The insoluble waste moves more slowly from its source, but in the course of time even the boulders and the sand, lubricated by rain water, slip and slide downhill and find their way into the rivers. Once there, they are moved along and are rubbed, rolled, and pounded together until they eventually find their way into lake basins and the sea.

Gullies developed by the erosion of a cliff in the Petrified Forest of Arizona, with a close view of the erosion pattern on a freshly exposed bank. (Gardner Collection, Harvard University.)

The rainfall in different parts of the world ranges from less than an inch up to more than 600 inches a year, and it is generally greatest where the mountains are highest and the slopes steepest. Here the action of rain is so powerful that the slopes are often washed clean of all waste, leaving great areas of bare, jagged rock. Indeed, if it were not for the presence of plants, especially grasses, which bind the soil together and hold it in place, there is no doubt that large areas of the earth's surface would soon be barren, rocky wilderness, entirely uninhabitable by man or beast. One can witness the erosive work of rain on any hillside that has been burned or deforested. Rain starts little runnels in the loose soil; these get deeper and deeper, to become gullies; and soon a network of gullies cuts the hillside, spreading in every direction, until all the soil is washed into the nearest river and only a barren, stony slope is left.

Rain works most easily on loose rock waste, but a river can cut its way into solid rock. How else could one account for the Grand Canyon of the Colorado, the gorge of the Zambesi, and countless instances on a smaller scale of rivers that are entrenched between solid walls of rock? The tools with which a river excavates its valley are the boulders and the sand that it sweeps along

Grand Canyon of the Colorado River. (Gardner Collection, Harvard University.)

Mature river with typical flat valley and meanders—Steamboat Springs, Colorado. (Sheldon Judson.)

with it. The constant bumping and rubbing of these materials on the river bed wear it down; and as the river surges from side to side, the walls become undermined, slabs of rock break off and fall into the river, and the gorge is widened as well as deepened. The action is slow by human standards, but it goes on year after year, century after century, aeon after aeon; and the result is the network of valleys that diversify the face of the earth and the stupendous quantity of waste that the rivers carry down to the sea.

It is reckoned that the discharge of rock waste by the Mississippi is equivalent to the lowering of the surface of its entire drainage basin by about a foot in every 4,000 years. It has also been calculated that if all the rivers in the world continued their work of abrading and removing at the same rate as now, they would reduce all the land of the world to sea level in some 10 million years.

There is reason to think that rivers have been running on the lands and carrying sediment to the sea for billions of years. How is it then that any land is left above sea level? Because the crust of the earth behaves like a seesaw; while one part is being worn down another part is rising up, and new land is constantly being formed.

In desert regions, where rain hardly ever falls, the task of engraving the land and transporting the material is taken up by wind. Most of the great deserts lie in the trade-wind belts of the Northern and Southern Hemispheres, where the wind blows from one quarter for a large part of the year. The trade winds are strong, dry winds; they parch the lands over which they blow, and only the hardiest of drought-resisting plants can live there. The rocks, deprived of shade, are exposed to the full heat of the tropical sun and are shattered relatively quickly, breaking down into coarse gravel and sand. The smaller sand grains are lifted by the wind; the bigger grains are rolled along the surface of the desert, and these moving grains abrade the surface of every rock and stone they come into contact with. The violence of this action can be measured by its results; in large parts of the Egyptian and Nubian deserts, and in the Namib or coastal desert of South-West Africa, every projecting rock and stone is grooved and polished on its windward side by the sand blast, while the

Sediment transported by wind, some held by vegetation—Scotland. (Gardner Collection, Harvard University.)

leeward side may be almost unaffected. The destructive action of sand-laden wind may be seen on the Pyramids and other ancient monuments near Cairo; yet it is recorded that inscriptions painted with red ochre on the leeward side of such monuments over fourteen centuries ago are still quite legible. Glass bottles left in the desert become opaque on the windward side in a few days and may be worn right through in a few months.

The popular impression of a desert is a surface entirely covered with loose sand, but this is quite wrong. The typical desert surface is one of bare, shattered rock and gravel, from which all the lighter grains have been swept away by the wind to form sand drifts against cliffs and in sheltered spots and hollows. On the leeward side of the desert, where more humid conditions prevail and grass can grow, the wind-blown sand is arrested and eventually bound together by grass. Large areas in China are covered with a deposit of very fine sand or dust which may be hundreds of feet thick; this has been carried by wind from the deserts of central Asia and caught and bound together by the vegetation of the grasslands. From west Africa, large quantities of sand and dust are blown out to sea.

It has not been possible to make any estimate of the rate at which the desert lands are being lowered by wind action, but deserts are generally rather low lying and it is not unlikely that wind is even more effective than water in lowering a region which is already of rather low relief and where the transporting power of water would consequently be at its lowest.

In the polar regions, and on high mountains, the work of running water is taken over by ice. When snow collects to a great thickness, it becomes compacted into granular ice, and when the mass is bulky enough it begins to spread and to flow downhill under its own weight. Valleys that intersect the highland are invaded from above by rivers of ice, or glaciers, which travel slowly downward at a rate of a few feet a day. A glacier carries with it, just as an ordinary river does, a load of rock waste, some of which is picked up by the bottom. Much of this stuff consists of sharp, angular boulders shed from rock surfaces that have been shattered

Sharp, angular fragments and boulders broken from rock surfaces by frost action—Selkirk Mountains, British Columbia. (Gardner Collection, Harvard University.)

by the action of frost. These angular lumps, embedded in ice and pressed down by the weight of an ice sheet perhaps one or two hundred feet thick, are admirable graving tools. Where they are dragged along the rocky floor, they cut deep grooves in it (and are, themselves, rounded and worn down). Finer waste grinds and polishes the floor. The best demonstration of the abrading power of a glacier is given by the streams that flow out of it when it melts away; these are very milky in appearance, owing to the quantity of fine rock flour that they carry along. It is sometimes possible, when a glacier has receded after a succession of warm seasons, to see part of its floor. This is always found to be deeply grooved or fluted in the direction in which the ice moved.

But if a glacier is a much more efficient carving agent than a river, it is much less efficient as a transporting agent on account of its slow rate of movement. It is certain that a glacier carries in a year a much smaller amount of waste than a river that drains an equal area of mountain and valley. The great ice sheets of Greenland and Antarctica deliver their load of rock waste directly to the sea, but valley glaciers like those of the Alps melt before they reach the plains and drop their load, which is carried on by rivers.

In most regions of the earth, a hole drilled into the ground encounters water before great depths have been penetrated. The water may appear only as moisture, or with a flow that comes out at the surface; but within reach of a pump it will often appear in quantities that can be removed at the rate of many gallons per minute. This is water that sank into the ground from rainfall, while other water evaporated back to the atmosphere or ran off along the surface in streams. Water plays an important part in the geologic drama, and supplies needs of civilization.

An aura of mystery has at times cloaked discussions of water underground. Aristotle supposed it to be gathered by the condensation on cavern walls of moisture evaporated from sea water carried under the lands by myriad streams fed from the sea bottom. For centuries, Aristotle's underground streams have been hunted for by men with forked sticks called divining rods, which have had a varied but always ignoble history. These same rods have been used to distinguish criminals from honest men, to determine the sex of unborn children, or to otherwise endow their operators with mysterious powers of divination.

Water moves from place to place underground, usually by percolating a few feet a year in closely packed materials or through slight cracks in rock. In coarse gravels, it can move many feet a day. But only in very special situations does it actually show as an underground stream. This is in areas where rock like limestone has been dissolved to such an extent that large, continuous channels were formed beneath the surface. In extreme cases, such channels become the underground caverns familiar to us all.

Many of the characteristics of water underground are discussed in the following selection by famed Harvard geochemist Prof. Robert M. Garrels, from his book **A Textbook of Geology.**

Robert M. Garrels

WATER UNDER THE GROUND

What becomes of the rain that does not evaporate back into the atmosphere or run off along the surface in streams? Where does it go after it sinks into the ground? If the earth is very old, is it possible that there is any room left for more water in the pores of the rocks? Or does this ground water *perform geological work comparable to that done by streams or waves?* . . .

The presence of wells and springs tells us that at least locally water does occur within rocks, as well as at the surface in the form of streams, lakes, swamps, and oceans. If a hole is dug into the ground, the sequence of observations is almost everywhere the same: first the hole passes down through soil or rock that may be damp or dry, but does not provide any free water that flows into the hole. Then a zone is reached in which the earth material becomes damper, and finally water begins to ooze into the hole. If enough time is allowed, the hole eventually fills with water up to the level at which the free flow first began. The depth to which it is necessary to dig to observe this phenomenon varies greatly; in some places it is necessary to dig down hundreds of feet, in others the *zone of saturation* is reached within less than a foot. This depth is clearly related to the climate, the topography of the land, and the nature of the earth materials in which the water is contained. In general, the lower the area, the smaller the

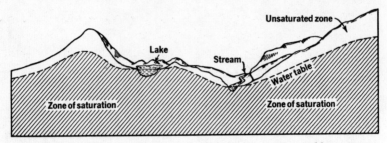

Cross section showing the relationship of the water table to topographic features.

distance to the zone of saturation. In humid climates, water is encountered relatively near the surface. Test holes show that any free water flow at the surface, such as a stream, represents an intersection of the zone of saturation with the ground surface. If enough holes are drilled, it can be shown that the upper limit of the zone of saturation, or *water table*, slopes upward away from streams or lakes, roughly paralleling the surface of the ground, but does not descend so abruptly as the valley walls or rise so high as the hill slopes. Consequently, the distance to the water table from the ground surface is greater beneath the crest of a hill than it is in the adjoining valley, but the zone of saturation can be reached in either place. It is rare that true underground streams are encountered; the water occurs in the pores of what is generally considered "solid rock." The few streams encountered are almost invariably in the relatively soluble rock called limestone, which is composed of calcium carbonate. In some extremely dense rocks such as granite, the rate of flow of the pore water is so slow that for all practical purposes there is no flow; in such rocks the movement of the water is controlled almost entirely by cracks or fissures in the rock.

The zone of saturation does not extend indefinitely downward; some wells and mines have penetrated below it, although the distance to dry rock usually is many thousands of feet. In general, we know very little about the lower boundary, for much of our information about subsurface materials comes from borings in search of water, and these stop when sufficient water is found.

In some cases, an interesting phenomenon occurs when borings are made; instead of the water slowly oozing into the drill hole, filling it up to the level of the first free flow, the water, after the zone of saturation is reached, rises in the hole into the zone of unsaturated rocks, and may even come up to the surface and overflow. Such wells are called *artesian*. They are of great economic interest; if the water comes all the way to the surface, the usual expensive pumping system is not necessary. Many cities in the Great Plains of the United States are served by such flowing wells.

If water is removed from a well that has penetrated the zone of saturation, one of two things may occur: either the rate of removal is less than the rate of replenishment, in which case the level in the well remains constant; or the rate of removal is greater than the rate of inflow, and the level of water in the hole continually falls.

Other phenomena are associated with ground water. Springs, for example, issue apparently from solid rock. The waters of springs may be hot or cold; the volume may vary from a mere trickle to many thousands of gallons a minute. They may issue from a point high on a valley wall, or from one just above stream level. The water may be almost pure, or may contain large quantities of dissolved salts. Some areas have many springs, some have none.

From these observations, it is clear that ground water must be able to move in most places, and that the controls of movement must be somewhat similar to those for streams. If this were not true, there would not be a fairly close correspondence of the

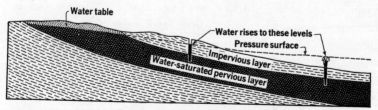

Typical condition for the occurrence of an artesian well.

topography of the ground water table to the topography of the land. The mere fact that the zone of saturation does not extend to the surface everywhere further attests to movement; in most places a few years' rains would saturate the ground from the present water table to the surface, if there were no underground movement. And, although movement occurs, most of the rocks of the earth must be strongly resistant to any erosive powers of ground water, because there is apparently little tendency to form underground streams except in the case of highly soluble rocks. Since these are the same rocks that are exposed to the work of surface water, it must be concluded that the energy of ground water is very small in relation to that of surface water.

Artesian wells indicate that some underground conditions exist that permit water to be trapped under pressure; if this were not so, the water would not rise above the point at which free flow was first observed.

Perhaps the most striking conclusion to be reached from the data is that the apparently solid rocks of the earth's crust are capable of holding and transmitting large quantities of water. Our first concern will be to see in greater detail what rock characteristics control the behavior of water within them.

The amount of water a given volume of rock can hold is determined by a property of the rock called porosity. Porosity is a measure of the empty space between the mineral grains which make up the rock. The space available in a given rock depends on many things: the shape and size of the grains, the way in which they are arranged or *packed*, and the amount of cementing material between them. For instance, a rock made of perfect spheres, without any cementing material in the spaces between them, would have a porosity of 48 per cent. . . . Such a rock is, of course, unknown, but the generalization can be made that the more uniform the grains, the larger will be the porosity. For instance, if an actual rock is composed of grains with a wide range of grain sizes, then there will be a chance that the smaller grains will fill the interstices between the larger ones. Many rocks are made of grains that are interlocking—the so-called *crystalline*

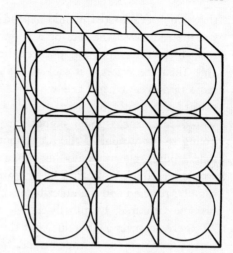

A sphere fills 52 per cent of the volume of a circumscribed cube. A rock composed of minerals packed as shown would have this maximum porosity.

rocks. Their porosity is usually very small, of the order of a few per cent. Essentially every gradation from complete interlock to the case of well-rounded, uniform, uncemented grains occurs in nature. The average porosity of natural rocks is probably of the order of magnitude of 10 per cent, if we consider only those exposed at the surface of the earth.

Although porosity controls the amount of water a rock can hold, a more important property from the standpoint of utilizing underground water is the ease with which water can move through them, called permeability. The rocks or materials that transmit water best are those in which the voids between grains are large, numerous, and connected.... The range of flow rates is almost incredible: on a standard basis for comparison a gravel deposit may allow a flow of several cubic feet per second, whereas a clay may transmit less than a millionth of a cubic foot per second....

Let us suppose for the moment that the earth is brand new, and that the surface is perfectly smooth and dry. Then, if we allow it to rain the same amount everywhere (not worrying about the source of such rain!) there would be no surface streams, for there would be no slopes, and the rain would either sink into the ground or evaporate. The amount that sank in, in comparison to that which evaporated, would depend on the permeability of the surface rocks and the rate of evaporation. If we further assume that all the rocks of the hypothetical earth are uniformly permeable, then the water that sank in would move downward at the same rate everywhere, and would continue to move downward toward the center of the earth. However, as it went deeper, it would enter rocks under higher and higher confining pressures owing to the weight of the rocks above. At no more than a few miles, pore spaces would be eliminated and the permeability would be reduced to zero. Thus a lower limit of movement would soon be reached and the outer layer of the earth gradually would become saturated. Eventually all the pore space would be filled (if there were enough rain) and the zone of saturation would extend everywhere to the surface. If the process went further, the earth would be covered by a shallow ocean.

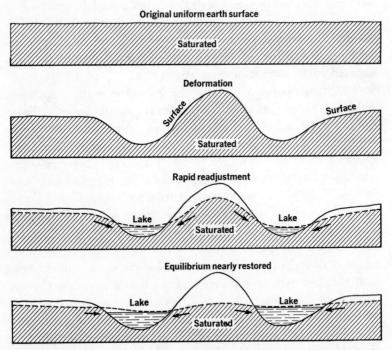

Predicted sequence of changes in the water table as a result of the deformation of a level, water-saturated section of a uniform earth.

If we were to interrupt this process before the pores all were full, and there still were an unsaturated zone near the surface, then any well dug anywhere on the surface would encounter the water table at the same depth, and the water table would be perfectly parallel to the surface.

What would be the effect if a mountain suddenly were formed on our hypothetical perfect earth? The surface of the water table would be warped; in the mountain it would have an upward bulge following the land surface, and in the adjoining trough it would be depressed beneath the general level. Water, even when semiconfined in porous rocks, still "tends to seek its own level"; so the water table in the trough would tend to rise back to the general level, while that beneath the mountain slopes would tend to sink. The result would be a general flow from the highs to the lows, and if the bottom of the trough were below the general level, the water would tend to collect in it as a lake or pond. Adjustment back to a uniform surface would be relatively rapid at first, and then would become slower and slower as the equilibrium surface was attained. The process would further be slowed by additional rain after the sudden formation of the bulge and trough; the rain would tend to maintain the high under the mountain. Thus we would have a system not unlike the conditions observed today: a distribution of water-saturated rocks conforming roughly to the topography, but more subdued because of the increased speed of flow when the water table is strongly warped.

Now, instead of having to form a hypothetical mountain, the same general sequence of events can be prophesied if there is a fairly extensive land surface being cut by streams. Let us assume that in the beginning the rocks are everywhere saturated with water to within 50 feet of the surface. As streams cut downward, their valley bottoms eventually intersect the water table; as they continue to cut deeper and to widen their valleys, a trough is cut down into the zone of saturation. This makes an outlet for the water in the rocks, which seeps into the channel from the saturated rocks along the sides. As the water drains out into the stream and is carried away, a local depression is formed in the water table, and the whole water level tends to be lowered. Of course, the effect is much more noticeable near the stream than farther away, because of the slowness of flow through the rock pores. The result should be just what is actually observed; the water table intersects the surface of the stream, and slopes upward and away from it. In fact, the over-all picture of the existence of ground water as observed fits very well a picture of a dynamic equilibrium between addition of water fairly uniformly in the form of rain, tending to saturate the rocks to a uniform level; and removal of that same water at varying rates in different places, because of the topography and because of the erosive processes at work. . . .

Some of the special ground water phenomena, such as springs and artesian wells, are explainable on the basis of earth materials of differing permeabilities. A spring may be defined as water issuing from the earth in a localized flow above the general level of the water table. Consider, for instance, an extensive land surface underlain by a permeable rock layer, in turn underlain by a highly impervious layer. As streams cut into the land surface, they intersect the water table, which may originally have stood near the top of the upper permeable layer. As the valley deepens, and the water table in the vicinity of the stream is lowered, the stream cuts down into the impervious layer. Flow from this layer is negligible, but flow continues from the pervious layer above. With each rain, water sinks into the ground, reaches the impervious layer, runs along the top of this layer, especially if it has a slight slope in the direction of the stream, and finally issues as a spring from the valley wall above the stream at the juncture of the impervious and pervious layers.

Many kinds of springs are recognized, but most represent a variation of this essential picture. Water movement may be controlled by one kind of rock or another, an open fissure may act in the place of the highly pervious layer, but in essence every spring depends on flow above the water table controlled by differences in permeability.

Artesian wells occur where a pervious layer of rock is covered by an impervious layer in such a way that the relatively free water in the pervious layer is trapped beneath the impervious one, and a hydrostatic head is developed. If a pervious layer of rock slopes downward from the surface and projects beneath an impervious one, the pervious layer admits water where it is exposed to rain. This water filters down rapidly into the pervious rock, and the layer fills up, much as a tilted pipe would fill. If, then, a drill hole pierces the capping layer and extends down to the pervious bed, the pressure is released and the water is free to move upward. Just how high the water will rise in the hole depends upon the height of the collecting area; if this is much higher than the land surface in the place where the layer is tapped, in all likelihood a flowing well will result. . . .

Caverns have been of great interest to man since prehistoric days, when they were used extensively for shelters. Imagination is fired by the eerie and grotesque rock formations that so often decorate the walls, and the strange forms of life inhabiting the perpetual darkness. In order to understand the processes at work, we will examine briefly some fundamental controls of the solution and deposition of limestone.

A piece of limestone can be placed in a vessel containing pure water and left there fifty years without losing a measurable amount of weight. But if a little of the gas carbon dioxide is bubbled through the water before the limestone is put in, we have quite a different result. Successive weighings at regular intervals

A typical limestone cavern with stalactites and stalagmites—Buchan Caves, Victoria, Australia. (Harvard University, Department of Geology.)

show that the limestone loses weight rapidly at first, then more slowly. If the weighings are continued long enough, it is found that the limestone finally reaches a constant lower weight.

From these observations, we can say that limestone dissolves in water containing carbon dioxide until the solution is saturated with limestone. The amount that dissolves in a gallon or two of water is not great compared to the solubility of many other substances, but the mere fact that it is possible to detect loss of weight in a limited amount of water indicates that the process may be important in nature, where great quantities of water are available.

Experiments also show that limestone dissolved in water can be recovered in a number of ways. One way is by evaporating the solution. When all the water has disappeared, a scum of limestone is left on the bottom and sides of the container. Careful work shows that the amount is just equal to the weight lost from the original chunk of limestone. Another way of recovering most of the limestone is to heat the saturated solution. Bubbles of carbon dioxide appear in the solution, and at the same time the limestone reappears as a finely divided solid that settles to the bottom of the container. . . .

As ground water containing carbon dioxide from the air percolates down through limestone, it dissolves the rock, rapidly at first, then more slowly until it is carrying a full load. Caverns are leached out of the rocks. Then, as time goes on, a change takes place. The solutions no longer continue to dissolve material when they reach the caverns; instead, as the drops of water seep down and emerge at the cavern roof, they evaporate partially and deposit some of their limestone. Gradually, a long "stonecicle" is formed. Such forms hanging from the ceiling are called stalactites. Below the stalactite, where the remainder of the drop falls to the floor, a complementary deposit rises from the floor. Such a mound or pillar is called a stalagmite. On the sides of the cave, where water runs down the wall, beautifully layered deposits are formed.

All the earth is made of elements in one form or another —solids, liquids, and gases. These constitute plants, animals, rocks, water, and atmosphere.

When the earth was in one of its early stages of development, it was a molten mixture of all the elements. As it began to solidify, many of the elements joined together and hardened into what we call minerals. Some of these joined to form the first rocks, called igneous rocks because they were produced by heat.

Igneous rocks (or others) that are exposed to weathering and erosion dissolve or break down into pieces that are carried to the sea and strewn on the bottom as sediment. This sediment eventually packs together and is cemented by elements deposited from solution, to form another class of rocks, called sedimentary rocks.

During the course of time, some rocks become buried deeply enough to be heated appreciably, and they undergo changes that obscure or destroy their original character; these constitute the third great class of rocks, called metamorphic rocks.

In looking at the varied rocks around us, we find that they all fall into one of these three broad classes. All are combinations of minerals, with about a dozen minerals dominant—and sands, gravels, clays, and soils are derived from rocks. Ninety-two elements have been found on the earth (a few more have been produced in the laboratory), but eight of these are so much more common than the others that they constitute over 98 per cent of the earth's crust. The materials of the earth are actually not so complicated after all.

The following article explains some of the characteristics and uses of elements, minerals, and rocks. It has been assembled from **Our Amazing Earth** by Carroll Lane Fenton, whom we met in connection with the sketch on Agassiz; **Mineralogy and Some of Its Applications** by Cornelius S. Hurlbut, Jr., Harvard mineralogist and author of **Manual of Mineralogy;** and **How to Know the Minerals and Rocks** by Richard M. Pearl, associate professor of geology at Colorado College and author of many popular books and articles on minerals, gems, and rocks.

*Carroll Lane Fenton, C. S. Hurlbut, Jr.,
Richard M. Pearl*

MATERIALS OF THE EARTH

Philosophers regard the earth as part of some universal scheme. Geologists look upon it as a rough ball of rock partly surrounded by water and wrapped in a covering of air. The petrologist thinks most about rocks and the minerals that compose them. Chemists care little for minerals or stone, but they constantly remind us that both are compounds of elements, which are the real building blocks of nature. They impress us mightily—until physicists turn elements into nuclei and electrons, which are the things of real importance, since they make the elements work. At this, many of us give up. We decide that it is fine to have an earth to live on, without puzzling our heads too much about its composition.

The fault is not ours, but our teachers! We have been told, more times than we can recall, that hydrogen and oxygen form water, while oxygen, nitrogen, and a few other gases make our atmosphere, or air. But who has told us that oxygen and silicon would make—if properly combined—the most durable of sandstones? Or that oxygen, united with carbon and some terrifically inflammable stuff called calcium, would produce the substance of limestone and marble? Such facts sound like bits from a geologic believe-it-or-not, designed to astonish the uninitiated. Actu-

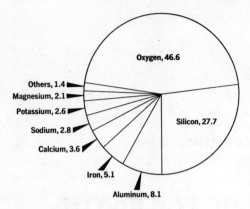

The relative abundance of the most common elements of the earth's crust. All figures indicate percentages. (Data from F. W. Clark and H. S. Washington, The Composition of the Earth's Crust, U.S. Geological Survey, Prof. Paper 127, 1924.)

ally, they are commonplace essentials. Without some knowledge of them we cannot understand the world.

Scientists have found that the world is built from atoms, of which there are 92 different kinds, and each different kind is called an element. Elements are the simplest substances known because, except by limited and unusual atom-smashing techniques, they cannot be taken apart or broken down into other substances.

Though over 90 elements exist in nature, only a few are important in minerals and rocks. Thousands of samples have been gathered at the surface, in the deepest mines, and at all possible levels between. Each sample has been analyzed chemically; from these analyses, tables have been prepared which show the percentage of each element in the earth's outer part, or "crust." The tables include all the known elements, yet only eight exceed 1 per cent of the whole, while only four exceed 5 per cent.

Elements combine to form minerals. In view of the age-old dependence of man on minerals for his weapons, his comforts, his adornments, and often for his pressing needs, it is surprising that many persons have only a vague idea about the nature of a mineral. Yet anyone who has climbed a mountain, walked on a sea beach, or worked in a garden has seen minerals in their natural occurrence. The rocks of the mountain, the sand on the beach, the soil in the garden are completely or in large part made up of minerals. Even more familiar in everyday experience are products made *from* minerals, for all articles of commerce that are inorganic, if not minerals themselves, are mineral in origin. All the common building materials such as steel, cement, brick, glass, and plaster had their origin in minerals.

Minerals are the products of natural processes. The greatest bulk of minerals occur as essential and integral constituents of rocks, others frequently are found in veins and cavities. Mineralogists limit the term "mineral" to include only those materials which occur naturally. Thus, steel, cement, plaster, and glass, although all are derived from naturally occurring mineral raw materials, are not regarded as minerals themselves, since they have been processed by man. Excluded, also, are all substances resulting directly from the processes of plant and animal life. Thus coal, oil, amber, and bones of animals are excluded, even though they occur naturally in the earth's crust.

The most important and significant limitation placed on the definition of a mineral is that it must be an element or a compound of elements that can be expressed by a chemical formula. This means that a specific mineral is always composed of the same elements in the same proportions.

If a mineral is composed of one element, its formula is merely the symbol of the element, as Au for gold. If the compositions of other minerals are to be expressed by formulas, the elements that make them up must always combine with one another in fixed simple ratios. For the common mineral quartz, the ratio is one atom of silicon to two of oxygen, with the formula SiO_2; and for the mineral pyrite, the formula FeS_2 indicates the ratio of one atom of iron to two of sulfur.

Now that we have determined what will be included and what excluded, we can frame a definition of a mineral as a naturally occurring element or compound formed as a product of inorganic processes.

Most of us have observed minerals in the process of growing. Anyone who has stood by a misted window on a chilly evening and watched the frost feathers spread their delicate fronds over

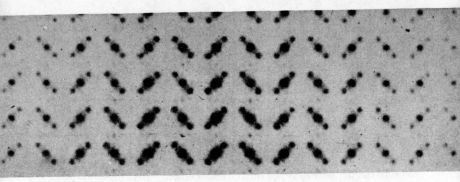

An X-ray photograph of atoms in the mineral pyrite, showing the orderly arrangement characteristic of crystalline structure. The mineral is composed of iron and sulfur, FeS_2. The large spots are atoms of iron; the small, atoms of sulfur. Each atom of iron is bonded to two atoms of sulfur, and the spacing of iron atoms is the same in both directions of the plane of the photograph. Magnification approximately 1.85 million diameters. (Martin J. Buerger.)

the pane has seen minerals growing in their natural environment; for ice is, by the most rigorous definition, a mineral. The water droplets, free a moment before to pursue with perfect mobility a winding course across the pane, are locked in an instant into the geometric rigidity of a spreading crystalline pattern. It is as though each tiny particle of water knew instinctively where to attach itself. If a small fragment of the frost is detached and placed in front of a photographic plate in the path of a beam of X rays, a symmetrical pattern of spots for each element and characteristic of crystalline solids will appear on the plate. A drop of the liquid water from the same windowpane so exposed will show no such pattern, indicating the absence in it of the orderly arrangement of its elements.

Further examination of the frost feather detached from the windowpane is rather disappointing, however, because we find that the restriction imposed by the flat surface and the mutual interference of the numerous rapidly growing crystals have so malformed the individual crystals that we can make out little about their symmetry. But if the process of crystallization of water vapor takes place high in the atmosphere where individual crystals are free to grow without mutual interference, the delicately beautiful and perfectly symmetrical snow crystals familiar to all are the result.

In a similar manner, every mineral has a natural habit of arranging its elements systematically into regular patterns, called crystalline structure. And if able to grow freely, it will, because of this natural habit of growth, develop into perfectly shaped crystals (external shape, determined by crystalline structure).

The mineralogist is being constantly called upon to identify minerals, and to this end he must bring to bear all the various means that he has at his disposal. If a mineral is well crystallized, the skilled mineralogist may be able to identify it by inspection or by the careful measurement of its crystal angles. However, such specimens are the exception; most minerals are found in imperfect crystals or irregular grains or aggregates.

Cubic crystals of pyrite which reflect in their shape the orderly arrangement of atoms shown in an X-ray photograph. The squares ruled on the background are 2 inches on a side. Striations that show clearly here are characteristic of this mineral. Note that those on adjacent faces are perpendicular to each other. The small specimen consists of three cubes interwoven. (Walter R. Fleischer, Harvard University News Service.)

To determine imperfectly shaped minerals, one must necessarily rely on other properties. The most important of these are physical characteristics such as color, luster, cleavage, hardness, and specific gravity. The color and luster can be observed at a glance. By a close inspection one can tell whether the mineral is transparent or opaque and whether it has the smooth, easy fracture called cleavage, or a rough, irregular habit of breaking. The touch is characteristic of such soft minerals as talc and graphite, which feel greasy, and taste is diagnostic for a few water-soluble minerals, such as halite (table salt). One can determine the hardness by touching the smooth surface of a mineral with a knife blade; and, finally, by lifting a mineral specimen, one can recognize at once whether it is light or heavy as compared with familiar substances of similar appearance and thus gain a rough idea of its specific gravity. These are all physical properties, but frequently they so uniquely characterize a mineral that they suffice for its identification.

Color. The color of a few minerals is so characteristic that it alone serves for identification. Anyone is quick to recognize the green of malachite, the azure blue of azurite, the pink of rhodochrosite, and the brass yellow of pyrite. In other minerals, however, the color varies from specimen to specimen. One must learn those minerals for which color is a distinguishing criterion: for example, the color in black quartz or green feldspar is not characteristic except at certain localities. On the other hand, pink is always characteristic of rhodochrosite. Red is diagnostic of hematite, but calcite can be any color.

Cleavage. Although the exterior of a mineral may be rough and irregular, the interior will have the regular, ordered arrangement of atoms characteristic of its crystalline structure. This atomic arrangement permits some minerals to break easily along one or more smooth surface. This property is known as cleavage. Its presence or absence is frequently an important aid in mineral identification.

MATERIALS OF THE EARTH

Mica cleavage, with 2-inch squares for size reference. Cleavage fragments lying beside the large block are of different thicknesses, as indicated by their different degrees of transparency. (Walter R. Fleischer, Harvard University News Service.)

Hardness. The forces acting to hold the atoms of minerals together vary greatly from mineral to mineral. If these forces are strong, only with difficulty can one pry the atoms apart by pressing on the surface with a needle point or knife blade, and the mineral is said to be hard. If the bonding forces are weak, the knife blade can easily be pressed into the mineral to scratch the surface. The difference in hardness exhibited by different minerals is easily determined in a qualitative way and can frequently be used as a determining criterion in mineral identification.

A mineral will scratch a substance of lower hardness rating than itself, and will be scratched by one greater in hardness. A useful practical scale for rough tests of hardness is:

1. Minerals that are scratched by the fingernail have a hardness of 2.5 or less.

2. Minerals that will not scratch a cent but are too hard to be scratched by the fingernail have a hardness between 2.5 and 3.5.

3. Minerals that will scratch a cent and are scratched by a good steel knife have a hardness between 3.5 and 5.5.

4. Minerals that are not scratched by a good steel knife are over 5.5 in hardness.

For example, the minerals apatite and beryl, both found in similar-appearing six-sided crystals, can easily be distinguished on the basis of hardness. Apatite can be readily scratched by a knife blade, whereas beryl, which is much harder, cannot be scratched.

Specific Gravity. The specific gravity of a mineral is the ratio of its weight to the weight of an equal volume of water. Determination of specific gravity is particularly helpful in the identification of some fine crystals or gem stones that one does not wish to mar in any way, as it involves merely weighing the specimen in air and again in water.

In general, specific gravity is dependent upon two factors: (1) the kind of atoms present, and (2) the packing of the atoms. Lead is a heavy element, and, consequently, minerals that contain lead will have a high specific gravity. On the other hand, aluminum is comparatively light, so we can expect aluminum minerals in general to have a relatively low specific gravity. However, two minerals made up of the same elements in the same proportions may have different specific gravities, for in one the atoms may be packed tightly together, while in the other they may be more loosely packed and a given number occupy a greater volume.

It has been pointed out that minerals are elements or compounds that have ordered internal atomic arrangements which give rise to certain physical properties. However, these properties may not be sufficient to determine the mineral. For such specimens, it is necessary to make chemical tests to identify the constituent elements of the mineral. Ordinary chemical procedure is often laborious, and the mineralogists have developed short cuts with the use of the blowpipe. However, it is impossible to obtain satisfactory tests for all the elements with the blowpipe, and recourse must often be had to the chemist's test tube.

Many chemical tests are qualitative; that is, they merely determine the presence of an element without regard to the amount. If all other tests fail, or if a new mineral is being described, it is necessary to have a quantitative chemical analysis in which the percentages of the constituent elements are determined.

People study minerals for different reasons: a petrographer specializes in the relatively small number of rock-forming minerals; mining geologists work with the ore minerals; crystallographers are concerned primarily with the crystal shapes of minerals. A mineralogist cuts across these limiting boundaries and studies all aspects of all minerals.

Methods of mineralogy are used every day to solve practical problems, and a person with mineralogical training finds many outlets for his talents outside of the field of pure mineralogy. The methods are applied successfully in many industries where the materials studied are not minerals, but synthetic products. For example, in the manufacture of abrasives, ceramics, refractories, synthetic crystals, and steel, mineralogical techniques have contributed much, and many of the manufacturers of these products employ full-time mineralogists on their staffs.

Mineralogy is an everyday tool with the mining geologist. Likewise, the prospector should be a student of mineralogy, for otherwise he can be led sadly astray. This is illustrated in the early history of Climax, Colorado.

Since 1913, the mines at Climax have produced a high percentage of the world's supply of molybdenum, mined as the sulfide, molybdenite. Molybdenum is used chiefly as an alloy to give greater strength and toughness to steel. Molybdenite is a black, soft, platy mineral and for 2,000 years had been confused with galena and graphite; these are also soft, black minerals, but contain no molybdenum. In 1890, mineral claims were staked on Bartlett Mountain at Climax, because of the presence of a black mineral which was thought to be galena. Galena frequently carries silver, so the prospectors thought their ore might yield both silver and lead. Some of the "galena" was sent to an assayer for analysis, but was determined to be a poor grade of graphite! As a result, the prospectors dropped their claims. It was not until five years later that the mineral was correctly identified as molybdenite. Had it been identified correctly at first, the ownership of the great mines at Climax would be different today.

The tests necessary to differentiate these minerals are simple, and anyone could perform them unerringly. Because of more general mineralogical knowledge, it is safe to say the mistake would not be made today. Galena can be told by its high specific gravity and three cleavages at right angles to each other. Molybdenite and graphite are both platy minerals but can be easily distinguished by specific gravity and streak. There are also several good blowpipe tests for molybdenite, whereas graphite is quite inert under the blowpipe.

There are certain mineral associations which are so characteristic that they may be as important in identification as color, cleavage, or hardness. For example, the most common minerals at Franklin, New Jersey, are black franklinite, greenish willemite, and red zincite in white calcite. This combination is so characteristic of Franklin that one can recognize all four minerals immediately.

Another example is the deposit of emery at Chester, Massachusetts. Emery, a mixture of two minerals, corundum and magnetite, has for centuries been used as an abrasive. For many years it was mined on islands in the Greek Archipelago and on the mainland of Turkey. In 1846, the Turkish government employed J. Lawrence Smith to study their deposits, and his report, including the mineralogy, was published in 1850 in the *American Journal of Science*.

In 1860, Dr. H. S. Lucas discovered at Chester, Massachusetts, a vein of black magnetic material which he believed to be magnetite. Magnetite, magnetic iron ore, was a well-known mineral and at that time was being mined at several places in the eastern United States. Attempts to smelt the Chester material as an iron ore met with failure; it did not react in the furnace as iron ore should. Accordingly, Charles T. Jackson, State Geologist of Massachusetts, was consulted. His examination of the property showed that the "iron ore" was associated with chlorite, margarite, diaspore, and chloritoid. These minerals in themselves meant little to him until he recalled that the exact association had been reported by J. Lawrence Smith in his paper on the Turkish emery. Thus, by association, Jackson identified the "iron ore" as emery, and, as a result, an abrasive industry sprang up in the small town of Chester. Although the Chester emery was mined out long ago, the manufacture of abrasive materials persists there to the present day. In 1860, emery was considered a distinct mineral species, and not until some time later was it discovered to be a mixture of magnetite and corundum.

Minerals combine to form the rocks of the world. Consequently, the difference between a rock and a mineral should be clearly understood. Rocks are the essential building materials of which the earth is constructed, whereas minerals are the individual substances that go to make up rocks. Thus granite (a rock) is composed of at least two minerals (quartz and feldspar), though others are almost certain to be present.

If a single mineral exists on a large enough scale, it may also be considered as a rock, because it may then be regarded as an integral part of the structure of the earth. Thus, a pure sandstone or quartzite rock contains only one mineral, quartz, distributed over a wide area. Other single minerals which are also regarded as rocks include dolomite, gypsum, serpentine, and sulfur—all of which occur in huge beds or masses. Some rocks of this type have a different name from that of the mineral composing them. Thus, the mineral halite makes rock salt; calcite is the constituent of the rock called limestone; and either calcite or dolomite can make up the rock called marble.

Each group of minerals is related naturally to definite types of rock. This enables us to identify the rock more readily than otherwise. Rocks are not as easy to distinguish as minerals because the different types grade imperceptibly into one another, but this principle of mineral association is very helpful.

Rocks include natural glass. Obsidian, an abundant rock in Mexico and Iceland, is natural volcanic glass. Organic products of the earth, which cannot be called minerals, are in some cases properly known as rocks. Coal, derived from partly decomposed vegetation, is a rock of this kind.

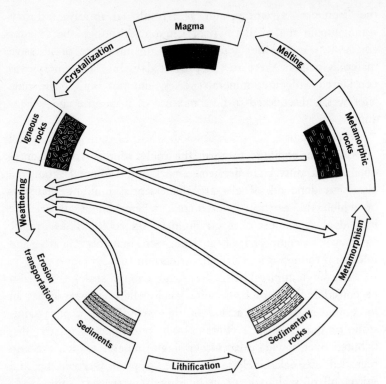

The rock cycle. Uninterrupted, the cycle will continue completely around the outer margin of the diagram from magma through igneous rocks, sediments, sedimentary rocks, metamorphic rocks, and back again to magma. The cycle may be interrupted, however, at various points along its course, and follow the path of one of the arrows in a short cut through the interior of the diagram. (From L. D. Leet and S. Judson, Physical Geology, *Prentice-Hall, Inc., Englewood Cliffs, N. J., 1958.)*

The many rocks which constitute the earth's crust are the result of geologic processes acting during long ages, building up some rocks and breaking down others. The normal rock cycle leads from molten rock to igneous rock, then to sediment and sedimentary rock, followed or preceded by a metamorphic stage. Countless bypaths to this cycle give rocks an infinite variety and prevent them from becoming monotonous to anyone who has gained an acquaintance with them and has a slight knowledge of geology.

All minerals and rocks have their primary origin in a body of molten rock called magma, which is believed to exist in local pockets deep within the crust of the earth. This magma eventually becomes the igneous rocks and minerals. The name "igneous," related to the word "ignite," suggests fire and heat.

Seismologists, the scientists who study earthquake waves, tell us that the earth beneath its relatively thin surface layers is not liquid, as was formerly thought, nor is there anywhere a complete zone of molten rock. Probably the hot rock is prevented from melting by the enormous pressure upon it, which maintains it in a semiplastic condition. When the pressure is relieved by cracks in the solid rock above, or heat due to radioactivity reaches the melting temperature, the rock slowly becomes molten and begins to rise.

As this magma comes to rest in a cooler place, but still within the earth's crust, it starts to solidify; thus the igneous rocks are born. They are known as intrusive rocks because they have intruded or forced their way into other rocks which were there already. This process has been going on ever since the beginning of geologic time; igneous rocks are presumably being formed in the same way today as they have been throughout the long history of our planet.

The common intrusive igneous rocks are granite, diorite, gabbro, and peridotite. Constituting the cores of mighty mountain chains, these rocks are revealed for observation only after millions of years of prolonged weathering and erosion by wind and rain and other agents of the atmosphere.

When the molten rock actually breaks through to the surface and wells out as a lava flow, or is blown out as volcanic fragments, the resulting igneous rock is called extrusive. We usually have in mind a volcano such as Vesuvius or Mauna Loa when we refer to this sort of igneous rock, but lava can issue quietly from open fissures in the earth without building a cone, as it still does in Iceland.

The common extrusive rocks are obsidian, rhyolite, andesite, and basalt. Formed upon the surface or at a shallow depth beneath a light covering, these rocks need not wait long before weathering and erosion sets in upon them.

Intrusive and extrusive igneous rocks differ chiefly because they have cooled at different rates. Intrusive rocks, losing their heat slowly while beneath the ground, acquire a coarse texture as the individual minerals have time to grow to a considerable size. Extrusive rocks cool rapidly; many grains get started, but each is small.

In extreme cases of sudden chilling, no minerals are visible at all, the only product being a natural glass. Obsidian Cliff in Yellowstone National Park, seen by a million tourists annually, is a world-famous example of volcanic glass which originated in this fashion.

Igneous rocks, such as granite and rhyolite, that are rich in silicon and aluminum tend to be light-colored and relatively light in weight. As the amount of aluminum is reduced, and the proportion of iron and magnesium increases, igneous rocks become darker and heavier, like gabbro and basalt.

The cooling of magma to form an igneous rock is accompanied by shrinkage and development of cracks called joints. Shrinkage also causes cavities or pockets, and these may later be filled or lined with crystals projecting toward the center. Some of the man-sized pockets of this sort yield large gemmy crystals of quartz, feldspar, and other minerals.

Another phase of igneous activity that concerns the mineral collector has to do with ore deposits. Metal-bearing solutions of many kinds accompany the rise of magmas. As the molten rock cools and becomes solid, large quantities of liquid and gas, charged with mineral matter, are given off. As these make their way slowly toward the surface, they may form mineral deposits wherever they encounter lower temperature, reduced pressure, or easily changed rocks.

During the long-distance migration of the solutions that have been expelled from the magma, ore deposits of gold, silver, lead,

Two geodes broken open to show their internal structure. Each is marked by an outer layer of dark silica, which is lined with milky-to-clear quartz crystals that project inward toward a hollow center. These structures are most commonly found in limestone, where they apparently form by the modification and enlargement of an original void. Two-inch squares on the background provide a scale. (Walter R. Fleischer, Harvard University News Service.)

zinc, and other metals are produced. These are referred to as veins because they run through the enclosing rock like veins within the skin.

Eventually, if not used up by one of the processes just described, the mineral matter that is left may appear at the surface of the earth in a volcano, gas vent, geyser, or hot spring. Around the volcanoes of the Mediterranean shore, for instance, are coatings of such minerals as native sulfur, realgar, and hematite. Amidst the spectacular fumaroles, or gas vents, of the Valley of Ten Thousand Smokes in Alaska are magnetite, pyrite, galena, and other minerals in large amounts. The geysers of New Zealand carry gold, silver, and mercury. Hot water at Steamboat Springs, Nevada, is depositing cinnabar today as in the past. There are numerous similar examples of each of these mineral occurrences, representing the final stages of igneous activity.

Even the most deeply buried igneous rock will some day be exposed by erosion. The age of the earth, several billion years as determined by the measurement of radioactivity in igneous rocks, allows time for very extensive erosion to have occurred almost everywhere. The forces of weathering will then begin to attack the rock, causing it to crumble and decompose. Some of the fine particles may be dissolved by rain water as it seeps through the soil and into the pores of the bedrock underneath. The rest may be washed away bodily by streams, or wafted by the wind, or carried in the frozen grip of giant glaciers.

When either the dissolved rock matter or the transported sediment is deposited somewhere else and hardens into firm rock, we have a sedimentary rock. Two types of sedimentary rock are possible, according to whether the original material had been dissolved in water or had been moved in the form of fragments.

In the first case—represented by such rocks as rock salt in Kansas and Michigan and beds of borax in Death Valley, California—the minerals are deposited when the dissolving power of the solution is reduced. This may happen because the water gets cooler or some of it evaporates or because of the action of certain plants and animals which extract chemicals from the water. In the same fashion sugar settles at the bottom of the cup when coffee cools, and salt incrusts the sides of a pan when salty water or brine is evaporated.

The second type of sedimentary rock is built up by the accumulation of separate grains of mud, sand, or gravel. Thus, mud becomes shale, sand becomes sandstone, and gravel becomes conglomerate. These sediments vary considerably in their mineral composition, and they grade into one another in the size of their particles.

Although the importance of wind and glaciers as transporting agents cannot be denied, most sedimentary material is nevertheless carried by streams. Rivers are therefore not only the great sculptors of the landscape and the chief creators of scenery; they play the major role in transporting the products of the earth that are to become the sedimentary minerals and rocks.

Probably the most intriguing sedimentary deposits are the ones known as placers, in which are concentrated gold, gems, and other heavy, durable minerals. The bearded Western prospector, equipped with gold pan and accompanied by his faithful burro, is the symbol of placer mining. Resisting chemical decay and physical damage alike, heavy minerals that will end in placers are washed from the higher elevations and taken downstream, until the force of the water is no longer sufficient to move them any farther. A slight obstruction in the channel or change in the current may be enough to cause them to drop to the bottom.

Billions of dollars' worth of native gold has been recovered from the bonanza placers of California, the Klondike, and elsewhere. Besides gold and a number of valuable gem stones—such as diamond, corundum (ruby and sapphire), spinel, and zircon—the most likely constituents of placers include magnetite, chromite, ilmenite, and cassiterite. Quartz, of course, is ever-present.

A special kind of placer is laid down along ocean beaches by waves and shore currents, which effectively separate the heavy minerals from the light ones. At Nome, Alaska, six beaches, two submerged and four elevated above sea level have yielded a good deal of gold in very tiny grains. Vast accumulations of ilmenite, rutile, and zircon line the beaches of India, Brazil, Australia, and Florida. So-called black sands, containing magnetite, ilmenite, and chromite, are extensively developed on the coasts of California, Oregon, and Japan. A most extraordinary representative of beach placers is the rich diamond-bearing bed near Alexander Bay in Namaqualand, South Africa, where diamonds brought down by the Orange River were distributed along the beach, in close association with oyster shells.

Wind often blows the smaller rock fragments into heaps called dunes. A single sample of dune sand may contain several dozen different minerals, but these are not of specimen interest except to collectors of sand, who must study them under magnification.

Owing to their bulk, glaciers are effective agents in transporting and depositing sediments. Unlike streams and wind, they are not selective in their action; a glacier embraces in its icy grasp boulders the size of a house, surrounded by particles that

Tillite, rocks hardened from the unconsolidated debris laid down by an ancient glacier a billion years ago. Scale is given by 2-inch squares in the background. (Walter R. Fleischer, Harvard University News Service.)

have been ground so relentlessly as to deserve the name "rock flour." With equal disregard for size, a glacier dumps the large and small material at the same time, with no attempt at sorting it. Such accumulations, common in every area glaciated during the recent ice age, are called moraines. Incorporated in moraines are minerals and rocks of foreign extraction which have been pushed, dragged, or carried bodily from their place of origin, in some cases hundreds of miles away. Chunks of native copper, brought down from the Upper Peninsula of Michigan, are strewn across the state of Wisconsin. Masses of chalcocite are frozen in a moraine at Kennecott, Alaska.

Loose sediment, whatever its origin, eventually becomes solid—"as hard as a rock"—because mineralized underground water cements together the individual grains and the weight of later sediments squeezes down upon them more and more tightly. At a fairly shallow depth, except in arid climates, the ground is saturated with water, which fills all pore spaces of the soil and bedrock. This water drains into streams or soaks out at the surface as seeps and springs. In caverns, stalactites hang from the roofs, while stalagmites build up from the floors—both the result of evaporation of underground water as it percolates into the earth.

The distinctive property of most sedimentary rocks is their

Geologists have found that these layers of sedimentary rock near Ithaca, New York were deposited 300 million years ago. (Gardner Collection, Harvard University.)

Metamorphic rocks are so changed by heat, pressure, and chemically active fluids that they have lost most of their original characteristics. (Gardner Collection, Harvard University.)

stratification, which refers to the layers of beds as each one is deposited on top of an earlier one. Just as the bottom book in a pile must have been the first one put down, so the lowest bed was the first one deposited, and each successive bed was formed at a later time. The heat of an igneous rock would consume any evidence of animal or plant life that might have existed, but many sedimentary rocks carry evidence of life.

Another characteristic of sedimentary rocks is the presence of foreign lumps or nodules called concretions. In the white cliffs of Dover are numerous odd-shaped pieces of flint, perhaps secreted by ancient sponges when the chalk that now makes up the cliffs was deposited in a shallow sea.

Joints are as abundant in sedimentary rocks as they are in igneous rocks, but the shrinkage which causes them is the result of drying instead of cooling. Most sediments are laid down in water—streams, lakes, the ocean—and may contain 50 per cent or more of moisture, some of which is driven off during burial.

The third main kind of rock, called metamorphic, is the result of drastic changes in igneous and sedimentary rocks. The rock has been changed so much from its original state that practically all signs of ancient life are gone, if there were any. Even most of the evidence as to the nature of the original rock has been lost, and often it is impossible to tell whether the pre-existing rock was igneous or sedimentary.

Heat from an invading magma that forces its way toward the surface of the earth is one of the factors that produces a metamorphic rock by creating new textures or entirely new minerals from the old rock. Another factor is pressure resulting from deep burial or slow movement in the earth's crust, pressure of the sort that ultimately bends rocks into mountain ranges. The chemical action of liquids and gases is also effective.

Limestone, for example, turns from a sedimentary rock into marble as the grains of calcite recrystallize under the influence of the agents of metamorphism. Because of the recrystallization, marble usually has a more glistening appearance than limestone. New minerals may be formed in the process, giving marble the swirled patterns that are so attractive a feature of colored marble.

All aspects of geology are brought to a focus in mountains. Mountains are the complex products of external and internal forces that deform and elevate rocks. These forces have produced what dry land there is and brought volcanoes into being.

Mountains are many in kind, location, and stages of development. In fact, to a geologist, all deformed and broken rock masses are mountain structures, past or present, even though they may not today stand above sea level. There have been mountains during every stage of earth history; mountains are being formed today in regions where earthquakes and volcanoes are most numerous.

Mountains go through cycles: sediments accumulate and harden to rock, are folded, elevated, eroded; the folding is renewed and erosion again takes over. They testify to mobility of the earth's surface, an endless restlessness in response to internal forces of origin as yet unknown.

The mountain story has been read from the rocks in great detail, but some of its mysteries are still to be solved. It is best told by the events of a single region where many of the features are illustrated. Such a region is the section of eastern North America known as the Appalachian Mountains, or Appalachian Highlands, the scene of the account that follows.

Half a billion years ago, sea basins were accumulating sediments that are found as rocks today in the Appalachians. The first folding and upheaval produced a range of Taconic mountains, then erosion destroyed them; Acadian mountains rose, only to be worn away. Volcanic activity as great as any known throughout the world produced countless volcanoes and flooded the White Mountain region with more than 10,000 feet of lava. In this area today stand mountains which are remnants of chunks of these lavas that foundered in the melting pot below them and were baked to resistant hardness.

The senior author of the article "Origin of the Appalachian Highlands" from which the following selection was adapted, Prof. Marland P. Billings, of Harvard University, is a past president of the Geological Society of America and one of the world's leading authorities on structural geology. The junior author, Dr. Charles R. Williams, of Liberty Mutual Insurance Co., is a mineralogist who has applied his training to protecting persons from industrial hazards such as silicosis. The article was originally published in **Appalachia.**

Marland P. Billings and Charles R. Williams

MOUNTAIN STORY

A mountain range is but a transitory creation. Like the gigantic redwood tree, it rises as a feeble sprout above the fertile earth; it grows, and in maturity lifts its lofty summit toward the sky. But it is doomed to destruction. The innate vigor which gave it growth slowly wanes and old age begins. The rain and the wind, the ice and the snow, the heat of midsummer sun and the frigid cold of winter night commence their work of annihilation. At first slowly, then more rapidly, the mountain range, like the tree, yields to forces of destruction, until at last only a barren stump remains as evidence of its former glory. But though the individual may die, the race endures, and a new entity arises where the old one fell. And so through countless eons mountain ranges follow one another in slow succession.

The Appalachians have had a varied history. Time and again their present site was flooded by shallow seas teeming with life. Not once, but several times, mountain ranges were uplifted in New England, only to succumb in turn to the forces of destruction. In the distant past, volcanic activity on a scale as stupendous as anywhere on the modern earth buried New England under lava flows thousands of feet thick. All but small remnants of once-extensive volcanic rocks have long since been swept to the sea. From volcanic heat, we turn to Arctic cold,

Every phase of geology is embodied in mountains—Lake Agnes and Mount Lefroy, Alberta. (Tozier.)

for in the recent past the northern Appalachians were buried beneath a mighty sheet of ice a mile or more in thickness.

Since the geologic history of the southern and central Appalachians was much simpler than that of the northern, it is natural that we turn southward for the first look at our problems. And since the structures become progressively more complex from west to east, we shall give our first attention to the Appalachian plateaus on the west, then the Valley and Ridge section, the Blue Ridge, and the Piedmont Plateau.

The Appalachian plateaus extend from central Alabama to the Mohawk Valley in New York State. From a width of 50 miles in the south, they gradually expand to a maximum width of 200 miles in western Pennsylvania and eastern Ohio. This is an exceedingly rugged region, cut to pieces by ramifying streams which branch again and again until they form an intricate system, called "dendritic" because it is similar to a tree and its branches. The relief—the difference in elevation between the valley floors and the hilltops—is frequently over 1,000 feet and some of the higher mountains reach 4,000 feet above sea level.

The Appalachian plateaus are comparatively simple geologically. They are composed of flat sandstones, shales. and conglomerates which range in age from Cambrian to Permian. The

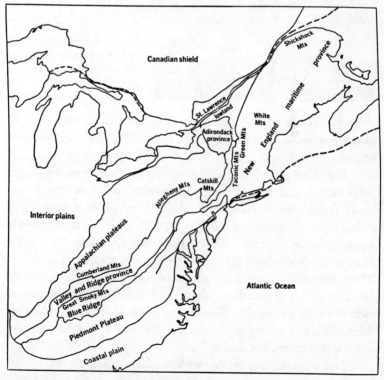

Subdivisions of the Appalachian Mountains and adjacent regions.

Dendritic drainage forms a pattern resembling the branching of trees— Africa. (U.S. Air Force.)

Pennsylvanian rocks of this region contain the vast deposits of coal which have been so important in the industrial development of the United States.

The Valley and Ridge region is vastly different from the Appalachian plateaus, in both topography and structure. It extends from central Alabama to Lake Champlain, nearly 1,000 miles, with its maximum width of 100 miles in central Pennsylvania. In sharp contrast to the Appalachian plateaus, the mountains of the Valley and Ridge region are long and narrow ridges. They frequently extend for scores of miles, although they may be only a mile wide. The valleys, and consequently the rivers, either run parallel to the ridges or cut across them at right angles. Accordingly, the drainage system resembles a gigantic lattice, and for this reason is called trellised.

This region supplies us with a classic example of folded

mountains. Gigantic waves, often several miles from crest to crest, extend for scores of miles. These folds have been deeply eroded, however, and the present mountains are but the resistant stumps of formerly great arches. From some of them not less than 4 miles of strata have been swept away, in part while the folds were being uplifted, in part during the vast eons of later geologic time. Folds which are curved upward are called anticlines; folds which are bent downward are called synclines.

This folded structure is characteristic of the northern Valley and Ridge region. In the southern section, the structure is different. Many of the anticlines have been broken by faults—dislocations caused by a slipping of rock masses along a plane of fracture. In some places, a whole series of anticlines have broken in this way and have piled up on one another like shingles on a roof.

The strata of the Valley and Ridge region are entirely of Paleozoic age. The Cambrian and Ordovician rocks are essentially the same throughout, but the overlying deposits vary. The Silurian and Devonian systems of the central Appalachians are 15,000 feet thick, but in Alabama they are only 200 feet thick, because a great delta was being built during the latter part of the Devonian in what we now call Pennsylvania.

The history of the Valley and Ridge region is relatively simple in its major features. Three principal stages have been recognized: (1) deposition of sediments during the Paleozoic (sedimentary rocks are almost invariably deposited in a horizontal position); (2) folding and faulting of these at the close of the

Syncline in a slate quarry in Pennsylvania. (Gardner Collection, Harvard University.)

Anticline in sandstones and shales, Hancock, Missouri. (Gardner Collection, Harvard University.)

Paleozoic; (3) erosion of tens of thousands of feet of the strata during and after the folding. Actually, many complications are shown in the details. Sedimentation was not continuous, but was interrupted by many intervals of emergence and erosion.

The Blue Ridge extends from Pennsylvania to Georgia. Its northern portion is only a few miles wide and 2,000 feet in elevation. It gradually increases in height and width toward the south, reaching its culmination in the Carolinas, Tennessee, and Georgia, where it expands to a width of 100 miles. Its highest summit is Mount Mitchell, loftiest peak east of the Mississippi River, with an elevation of 6,684 feet.

The northern Blue Ridge is composed dominantly of lavas, granites, and metamorphic rocks of Pre-Cambrian age, with quartzites of Cambrian age, whereas the southern part is underlain by metamorphic rocks and granites, mainly of Pre-Cambrian age. In the Great Smoky Mountains there is a tremendously thick group of unfossiliferous, strongly folded and metamorphosed sandstones, slates, and conglomerates of late Pre-Cambrian or earliest Paleozoic age.

The structure of the Blue Ridge is complex. The strata are so strongly folded and altered that it is impossible to trace individual folds for any distance. But the relationship of the Blue Ridge to the Valley and Ridge region is clear. Along most of its length, the former has been driven westward over the latter, in places a score of miles.

The Piedmont Plateau, as its name (foot of the mountain) implies, is not mountainous. Its relief never exceeds a few hundred feet and the absolute elevation ranges from sea level at New York City to 2,000 feet in northwestern Georgia. The drainage is dendritic, similar to that of the Appalachian plateaus; the geology is extremely complex.

Certain major phases of the history of the southern and central Appalachians are obvious. We have already noted that the Appalachian plateaus are composed of flat strata ranging in age from Cambrian to Permian. It follows, therefore, that throughout Paleozoic time the present site of the plateaus was dominantly an area of sedimentation. Many of the sediments were deposited in a sea, for they contain marine fossils. Similarly, the Valley and Ridge region must have been a region of heavy deposition

Layers of rocks that were once sandstones lying horizontally are now metamorphosed to quartzites and standing nearly "on end" in the Great Smoky Mountains. (L. B. Gillett.)

throughout Paleozoic time, for strata 25,000 feet thick still remain. Such regions of unusual sedimentation are called geosynclines. Although the sediments in the Appalachian geosynclines were many miles thick, the sea in which the formations were deposited was always shallow—never more than a few hundred feet deep. We must visualize a shallow sea, the floor of which gradually sank, with sediments accumulating during the sinking. The geosynclines were by no means continuously flooded by the sea. Frequently they were low plains standing a few feet above sea level, and much of the material was deposited in deltas. The greatest of these deltas was forming during the upper Devonian in central Pennsylvania, where strata 9,000 feet thick accumulated. During the Pennsylvanian the whole of eastern North America was a low flat country covered by vast swamps in which accumulated huge quantities of peat, the progenitor of our modern coal.

Whence came these sediments? The interior of North America could not have been their source, for it was itself frequently flooded by shallow seas at the same time that deposition was going on in the geosynclines. There are several lines of evidence which demonstrate that the sediments came from the southeast. The strata of any given age thin toward the northwest, and it is self-evident that any group of sediments will thin away from the source of supply. Even more compelling is the character of the materials across the geosynclines. The rocks become coarser toward the southeast. Formations of Devonian age, which are limestones in the Appalachian plateaus, become shaly toward the southeast and finally grade into sandstones. The Pennsylvanian formations become conglomeratic toward the southeast.

There must have been land masses to the southeast which were constantly rising throughout Paleozoic time and supplying sediments to the Appalachian geosynclines. They no longer exist, but in the products of their destruction we read the story of their life. The tremendous volume of the sediments—thousands of cubic miles—shows that these sources must have extended far out into the present Atlantic, at least several hundred miles.

SUMMARY OF EVENTS IN THE APPALACHIAN MOUNTAINS

Eras	Eras began, years ago	Periods	History of the Appalachian region
Cenozoic		Pleistocene	Glaciation of northern section
		Pliocene	Somerville peneplane completed; followed by uplift
		Miocene	Harrisburg peneplane completed; followed by uplift
		Oligocene	Kittatiny peneplane completed; followed by uplift
		Eocene	Erosion develops Kittatiny peneplane
	60 million	Paleocene	Erosion begins to develop Kittatiny peneplane
Mesozoic		Cretaceous	Fall Line peneplane completed; followed by uplift
		Jurassic	Erosion developing Fall Line peneplane
	200 million	Triassic	Red beds in basins; volcanoes; Palisades sill, fault-block mountains
Paleozoic		Permian	New England volcanism; White Mountains batholith; glaciation around Boston; Appalachian revolution; granites intruded
		Pennsylvanian	Deposition of sediments, including coal beds
		Mississippian	Acadian mountains destroyed, sedimentation southern part
		Devonian	Acadian revolution
		Silurian	First Taconic mountains destroyed; volcanism in Maine
		Ordovician	Sedimentation; Taconic revolution at close of period
	500 million	Cambrian	General sedimentation
Pre-Cambrian			Basement complex produced by sedimentation, volcanic activity, and igneous intrusions.

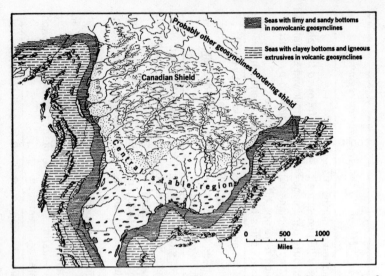

North America in early Ordovician time, showing island arcs and two belts of seas marking zones of geosynclinal deposition bordering a central stable region which was itself invaded repeatedly by shallow seas during the Paleozoic, but did not become involved in mountain-making deformations. (After Marshall Kay, North American Geosynclines, *Geological Society of America, Memoir 48, 1951.)*

The Paleozoic closed with a great climax of mountain-making activity. This brought about such a radical change of conditions that it is sometimes referred to as a revolution. Unable to withstand any longer great forces that had been developing in the crust, rocks southeast of the geosynclines were driven northwestward hundreds of miles. The sediments of the geosyncline, caught between the jaws of a great vise—the relatively passive interior of North America on the northwest, the moving masses on the southeast—crumpled into giant folds. Some rocks in Tennessee were so thin that instead of arching into folds they broke into many sheets, and sheet after sheet was driven northwestward along great thrust faults.[1]

More or less contemporaneously with the folding, great

---

[1] Faults which result from compression are called thrust faults; faults which result from stretching are called normal or gravity faults.

masses of granite were injected into the crust where the Piedmont Plateau now lies. Most geologists believe that the granites were a product of the same stresses which caused the folding. However, they frankly admit ignorance as to the ultimate cause of the great compression.

How high were these mountains during the late Paleozoic? We must confess that we have no direct evidence. But, if we reconstruct the folds of central Pennsylvania, we find that the crests of some of the arches rise 20,000 feet or more above the sea. If we then make allowance for strata which must have once existed, but which have been completely eroded away from the whole Appalachian region, we can safely say that some of these arches could have reached an elevation of 30,000 feet. However,

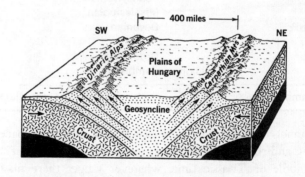

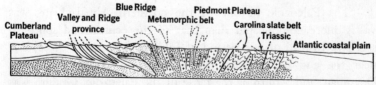

The top sketch illustrates a theory of the connection between a geosyncline and the mountains formed from it as applied to the Dinaric Alps and the Carpathian Mountains. The bottom sketch shows an interpretation of the symmetry of the Appalachians (width of section 400 miles), which is similar. (Top: After Leopold Kober, Der Bau der Erde, Berlin, 1921. Bottom: After Philip B. King, The Tectonics of Middle North America, Princeton University Press, Princeton, N.J., 1951.)

MOUNTAIN-FOLDING CALENDAR

Eras	Eras began, years ago	Periods	Mountain-folding episodes
Cenozoic	60 million	Pleistocene Pliocene Miocene Oligocene Eocene Paleocene	Himalayas Alps
Mesozoic	200 million	Cretaceous Jurassic Triassic	Rocky Mountains
Paleozoic	500 million	Permian Pennsylvanian Mississippian Devonian Silurian Ordovician Cambrian	Appalachians
Pre-Cambrian			Laurentians

we cannot disprove the possibility that erosion kept pace with folding and that no mountains ever existed! In any event, before the middle of Triassic time, whatever ranges there may have been were worn down to a low, rolling plain of slight relief.

The Paleozoic history of the northern Appalachians was vastly more complicated than that of the central and southern sections. Instead of one period of folding, there were three; one near the end of the Ordovician, a second during the upper Devonian, and a third at the close of the Paleozoic. Granite was injected into the crust several times, and volcanic activity complicated the problem. Many of the formations became severely metamorphosed and rocks of greatly differing ages and origins now appear to be similar.

Perhaps the most striking contrast between the surface appearance of the northern Appalachians and the district farther south is the abundance of lakes in the former. Where is there a lake more beautiful than Winnipesaukee or Champlain? Who can forget the solitude of Moosehead or of Rangeley? Who does not love the placid Adirondack lakes, nestled among their guardian hills? These enchanting lakes are a heritage of the ice age.

The White Mountains of New Hampshire are the most conspicuous topographic features of the northern Appalachians. Although for convenience they are referred to as ranges, they actually have no common trend, and in many cases their rocks are quite different. The highest peak, Mount Washington, attains an elevation of 6,288 feet.

Two major groups of rocks comprise the White Mountains of New Hampshire, an older group of metamorphic rocks and granites, probably Ordovician to Devonian in age, and a younger group of lava flows, granites, and associated rocks, probably of late Permian age. The metamorphic rocks were originally shales, sandstones, and limestones which were later folded, metamorphosed and then injected by granite. The relationship between the metamorphic rocks and the granite is complex.

The younger granite and associated rocks form a great body which extends many miles down into the crust of the earth and is known as a batholith. These rocks have congealed from magma many thousands of feet below the surface and have been exposed by erosion. In the White Mountains they are far more abundant

A rock that was formed when granite was injected into previously folded sedimentary formations. The white streaks are granite that soaked into the dark rock along its surfaces of easy breaking. (Gardner Collection, Harvard University.)

than lava flows. Within the limits of the main batholith, called the White Mountain batholith, lava flows and associated ash showers compose several of the mountains. Beyond the confines of the main batholith, the lavas are typically developed in the Ossipee mountains.

When magma came within a few miles of the surface, its fury could no longer be restrained and it broke through, giving rise to extensive volcanism. The whole White Mountain region was buried under floods of lava and showers of ash. These attained a minimum thickness of 10,000 feet, and probably much more. As the magma in the great chamber below continued to work upward, the roof of the batholith became so thin that great blocks of lava, from 5 to 9 miles in diameter, sank into the molten lake below. Each of the mountains composed of lava flows is one of these great blocks which subsided 1 or 2 miles into the magma, where it was thoroughly baked and toughened. In the meantime, the magma had spent its energy and it gradually cooled. Since those days of intense volcanic activity in the late Permian, many thousands of feet of strata, probably miles, have been eroded from the White Mountain region. The present ranges are but the hardened and resistant stumps of older mountains.

The Paleozoic formations of New England were deposited on a floor of Pre-Cambrian rocks which had itself been through a long and involved history. By early Cambrian time, the basement rocks had been reduced to a surface of low relief, so that when the Paleozoic seas first invaded New England they rapidly flooded most of the region. The Ordovician history is rather obscure, with seas covering western New England and probably extending as far east as the White Mountains. As in the southern and central Appalachians, these sediments must have come from land to the southeast.

At the close of the Ordovician, western New England was subjected to strong horizontal compression, and the Cambrian and Ordovician strata west of the present Green Mountains were intensely folded. This period of compression has been called the Taconic revolution, for the strata of the Taconic mountains were first folded at that time.

Horizontal Cambrian layers on an eroded surface underlain by vertical Pre-Cambrian beds that had been through a long history of mountain-making—Box Canyon, Ouray, Colorado. (Kirtley F. Mather.)

After the Taconic revolution, rapid erosion set in and by the beginning of the Silurian the mountains were gone. New England, however, was not inundated again until the middle Silurian. In northeastern Maine, intense volcanism occurred, great floods of lava and showers of ash covering the region. During lower Devonian time much of New England east of the Connecticut River was flooded alternately by marine waters and volcanic rocks.

With the advent of the middle Devonian, the Acadian revolution began. The great land masses to the southeast began to migrate toward the northwest and for the second time during the Paleozoic the sedimentary rocks of New England were squeezed from the side. The strata were folded into great anticlines and synclines and several thrust faults developed. It is probable that the rocks of the whole New England–Maritime region were folded at this time. Then with the Acadian revolution, the sea was driven from New England forever.

The later Paleozoic history of much of New England is obscure. It is probable that there was a long interval of erosion during the Mississippian and the early Pennsylvanian. In the middle of the Pennsylvanian the land which lay to the east of Cape Cod began to rise, and several basins of deposition formed in eastern New England. As far as we know, however, none of these basins were connected with the ocean, and all the deposits seem to be continental. Many of the materials were very coarse conglomerates. Mississippian and Pennsylvanian deposition was extensive. During the Permian, a great glacier came down from the eastern mountains to deposit the Squantum tillite[2] near Boston.

After the deposition of the Mississippian and Pennsylvanian rocks, land on the southeast once more crowded toward the northwest, jamming the strata into great folds. In many localities the compression was so great that folding could not accommodate the tremendous shortening, and numerous thrust faults developed. This period of folding affected only the eastern part of New England, but was as great as the earth has known. At about this same time, an almost equally intense mountain-making movement affected central Europe and southern Asia, forming mountain ranges from Ireland to southern China.

Contemporaneous with all this folding at the close of the

---

[2] Tillite is rock hardened from unsorted fragments deposited directly from glacier ice. (When it is not hardened into rock, such glacial debris is called till.)

Paleozoic, which has been called the Appalachian revolution, or shortly thereafter, great masses of molten granite were injected into the rocks of eastern New England and the Piedmont Plateau of the southern Appalachians.

But the Appalachian Mountains of the late Paleozoic, like all their predecessors and their successors, were doomed to destruction. By the middle of Triassic time, they had been worn away, and in their stead a low featureless plain pervaded eastern North America.

The crust was still restless, however; some parts of the great plain began to rise, other parts began to sink, erosion attacked the uplifted portions and the products of their destruction were swept into the basins. As the highlands rose more and more, faults began to break up the crust, and fault-block mountains developed. The topography must have been very similar to that existing today in the Great Basin of Nevada.

The three largest of these basins are still preserved around the Bay of Fundy, in the Connecticut Valley, and in a long belt extending from north of New York City to Virginia. Minor basins are found farther south. The Triassic rocks of these basins contain large amounts of red beds, such as red sandstones, conglomerates, and shales. The red coloring implies a semiarid climate throughout part of the year. Volcanic activity at this time formed the Palisades of the Hudson, a great sill—a body of magma which was injected parallel to the bedding. As the magma cooled, it contracted into the six-sided columns which characterize the Palisades. After many thousands of feet of sediment had been deposited in these basins, additional faulting occurred, chopping up the basin into a complex mosaic. Then, once again, the powerful forces of erosion gained the upper hand and the process of reducing eastern North America to a plain of low relief recommenced.

By early Cretaceous time the streams had reduced it to a surface that rose only a few score feet above sea level. Such a surface is called a peneplane—"almost a plane"—and this particular one is called the Fall Line peneplane. With the completion of this surface, a new phase in Appalachian evolution began. It will be recalled that during the Paleozoic the Appalachians, particularly the northern part, were repeatedly subjected to horizontal compression. During the Triassic, some of the broadest arches were so stretched at their crests that gravity faults and fault-block mountains developed, but with the completion of the Fall Line peneplane a third kind of stress became dominant: vertical forces began to uplift the whole Appalachian region, operating throughout the later Mesozoic and Cenozoic. The uplift has been erratic, some areas being pushed up hundreds or even thousands of feet higher than others. The rate has also varied greatly. For short periods the crust has risen rapidly and then for long intervals stood still. During periods of quiescence the rivers have been able to develop wide, flat valleys or partial peneplanes.

After completion of the Fall Line peneplane in early Cretaceous time, the whole of eastern North America was uplifted and subjected to a long interval of erosion. By early Cenozoic, streams had reduced the Appalachian region to a new surface of low relief, the Kittatiny peneplane. Above this surface, certain isolated mountains remained, either because of their superior resistance to erosion or because of their location on drainage divides. Mount Monadnock in New Hampshire is the type example of such residuals, hence all similarly isolated mountains rising above a peneplane are called monadnocks. In some localities, mountain ranges were left standing above the general surface of low relief.

As soon as eastern North America was uplifted after the completion of the Kittatiny peneplane, the streams had steeper gradients and at once began to cut into the old surface. Large parts of the peneplane were thus rapidly destroyed, but even today small remnants remain as flat surfaces at high elevations, witnesses to its former all-pervasive character.

In the middle of the Cenozoic there was a period of crustal stability during which most of eastern North America—perhaps 50 per cent—was reduced to a new surface of low relief, the

Harrisburg peneplane, so named because of its typical development around the capital of Pennsylvania.

But long before erosion could reduce eastern North America to a surface as widespread as the older Kittatiny peneplane, renewed uplift gave the streams additional power and the cycle of erosion began again. In the later Cenozoic there was another interval of crustal stability, and a new peneplane, the Somerville, was developed. This surface was not as extensive as the older peneplanes. Finally, not much over a million years ago, vertical forces uplifted eastern North America once again and the streams began to cut into the Somerville surface, where they are working today.

With the advent of the Pleistocene, Appalachian evolution entered a new phase: the Great Ice Age. The atmosphere slowly cooled. In the higher ranges of the northern Appalachians the winter snows lingered longer and longer into the summer months. Eventually, some of the snow never melted; it gradually accumulated until it became compacted into ice. Mountain glaciers thus developed and carved out great amphitheaters, or cirques, on the flanks of the White Mountains, the Adirondacks, and Katahdin.

Meanwhile, great icecaps had been accumulating around Hudson Bay. As the ice became thicker, it gradually spread out in all directions, just as a mass of sticky molasses would slowly spread. As the great ice sheet moved southward into New England, its advance was blocked by the White Mountains and the glacier was temporarily stalemated. For a while it awaited reinforcements from the north and then with renewed strength it poured through the notches, scouring and cleaning them of all projecting spurs. But the ice to the north was growing thicker, and finally the mountains were overwhelmed by the flood of ice.

The Green Mountains and the Adirondacks were also submerged beneath the sea of ice. Once these great mountain barriers of the northern Appalachians had been overcome the glacial front slowly pushed southward and in the heyday of its existence it extended as far south as Long Island and New York City. At this time, New England was buried beneath a mile or more of ice. Then tropic heat became victorious over Arctic cold. The ice began to melt away and the front was slowly driven northward. The highest peaks began to emerge through the shrinking cover of ice. With time the main ice sheet completely disappeared and only in the highest ranges small and inconspicuous mountain glaciers lingered on. But they, too, have long since passed away.

In this discussion, we implied that just one great continental ice sheet formed, swept southward over the northern Appalachians, and then melted away. However, evidence indicates that the glacial cycle was repeated four times during the Pleistocene's million years.

The interval since the melting of the last icecap is insignificant geologically—a mere 25,000 years. Still, many of the glacial deposits have been swept away and locally deep gorges have been cut in bedrock. Glaciation left the northern Appalachians with a disrupted drainage system and countless lakes and swamps. But the lakes and the swamps will be slowly destroyed, both by sedimentation and by the gradual lowering of their outlets.